Introduction to Research
Multiple Strategies for Health and Human Services

D1244461

Elizabeth DePoy, Ph.D., M.S.W., O.T.R.
Associate Professor
School of Social Work
University of Maine
Orono, Maine

Laura N. Gitlin, Ph.D.
Associate Professor
Department of Occupational Therapy
Assistant Director
Director of Research
Center for Collaborative Research
College of Allied Health Sciences
Thomas Jefferson University
Philadelphia, Pennsylvania

 **Mosby**

St. Louis Baltimore Boston Chicago London Philadelphia Sydney Toronto

Mosby

Dedicated to Publishing Excellence

Sponsoring Editor: Jim Shanahan
Production Manager: Nancy C. Baker
Project Supervisor: George Mary Gardner
Proofroom Manager: Barbara M. Kelly

2   3   4   5   6   7   8   9   0   GW/MA   97   96   95   94

Library of Congress Cataloging-in-Publication Data
DePoy, Elizabeth.
    Introduction to research: Multiple strategies for health and human
services/Elizabeth DePoy, Laura N. Gitlin.
       p.   cm.
    Includes bibliographical references.
    ISBN 0-8016-6284-2
    1. Public health—Research—Methodology.   2. Medical care—
Research—Methodology.   3. Human services—Research—Methodology.
4. Experimental design.   I. Gitlin, Laura N., 1952-   .   II. Title.
    [DNLM: 1. Research Design.   W 20.5 D422i   1993]
RA440.85.D47 1993
610'.72—dc20
DNLM/DLC   93-9724
for Library of Congress   CIP

To Don and Sam
Edward, Eric, and Keith
and our students

FOREWORD

This book is about the conduct of human inquiry, about people doing work with other people, about making sense of human experiences, thoughts, and actions.

In this comprehensive volume Elizabeth DePoy and Laura N. Gitlin provide an integrating framework not only for the conduct of naturalistic and experimental inquiry but also for the relationship of theory and practice. These experienced professionals and researchers describe *ways of knowing* about health and illness, about people who are embedded in contexts of delivering health and human service. Unlike other textbooks, their work is integrative. It recognizes that each way of knowing is incomplete, fragmentary, partial. Consequently, because of this, any single methodology is at the same time both trustworthy and untrustworthy in what it helps people to illuminate and understand. It recognizes that each way of knowing is not independent of the other, not an *alternative to* but *complementary with* other ways of knowing. The ideas of complementariness and connectedness are celebrated rather than bemoaned, legitimized rather than delegitimized, as is too often the case. Because DePoy and Gitlin allow these values to explicitly influence their writing, they avoid presenting material and arguments in smorgasbord fashion. And more important, in an intellectual sense, they are able to write about and illustrate methodology at multiple levels—political, philosophical, technical, cognitive, and linguistic. Herein lies a unique and important strength of their treatment. DePoy and Gitlin recognize that the act of doing research as well as the product of that act are a complex blend of political, ethical, technological, social, and linguistic engagements between human beings. Both the structure and content of their text reflect this idea.

DePoy and Gitlin have benefited from the methodological debates characteristic of the 1970s through the early 1990s. Their work, both at theoretical and practical levels, articulates the important dimensions that characterize, in their words, "the continuum of experimental-type design" and the "continuum of naturalistic inquiry." However, their project reaches beyond the boundary of research continua to integrate the fragmentation or discontinuities often generated by the actors in these debates. Thus their text reflects an interest for integrated approaches to inquiry. And just what is integrated? And at what levels does this integration take place? Ideas of what constitutes sound inquiry in one tradition (e.g., ethnography) are related to those of another tradition (quasi-experimental). In turn, how one goes about this integrated work and how one writes up such integrated work are described. Thus the language, cognitive processes, and value systems of those engaged in integrative designs at different levels are carefully articulated throughout the text.

Those who do research in health care settings will find a great deal of practical knowledge grounded in a sophisticated philosophical and sociological context. This should be refreshing. Those who are about to embark on doing research will not only learn how to do good work but, and perhaps more important, understand why they do what they do. Consistent with the viewpoint taken by DePoy and Gitlin, their text ends by empowering those who do the work. "Stories From the Field," the final chapter, is a powerful and constructive narrative way to know about the conduct of research.

John D. Engel, Ph.D.
Director, Program In Medical Humanities & Social Sciences
   and Research Associate Professor of Family Medicine
Jefferson Medical College
Philadelphia, Pennsylvania

PREFACE

Our main purpose in writing this book is to share with the student, helping professional, and beginning researcher our enthusiasm and passion for conducting research in health and human services. We hope this book demystifies the research process and provides helping professionals with a foundation from which to critique and understand the research designs and their applications to health care and human service settings.

You may well ask, Why another research text? The health and human service professions stand at the crossroads of many important trends and changes in the delivery and financing of services to clients, their families, and the community at large. Underlying these trends is the growing recognition of the importance of research to develop, refine, and validate the knowledge base from which effective professional practices develop. Helping professionals have increasing demands on their time to either initiate research projects, participate as members of interdisciplinary research teams, or to critically understand research articles and findings in their respective professional journals. Also, new directions in health and human services research indicate that the traditional scientific research paradigm, referred to as the rationalistic approach, quantitative approach, or experimental-type research that most students are taught, represents only one appropriate way of developing and testing knowledge. Another emerging approach is referred to as discovery-oriented, qualitative, or naturalistic inquiry. The historic opposition of quantitative-oriented methodologies to those that yield qualitative information has limited the professional's understanding and effective use of each approach. The ongoing debate between the two schools of thought has encouraged the idea that one method is more valid, correct, or scientific than the other. Rather than participate in the philosophic debate, many researchers currently encourage the use of multiple methodologies and the integration of approaches to understand the diverse realities of client groups, health care, and human services.

Students, professionals and researchers need a text to prepare them for these exciting transformations in research in the health and human service fields. This book provides the reader with a comprehensive understanding of how researchers think and act in both naturalistic and experimental-type social and behavioral research and how such distinct design strategies can be integrated to advance the level of knowledge that one obtains. The reader will learn how to critically evaluate, implement and respect each research strategy from its own philosophic perspective, thinking process, and specific actions that engage the researcher.

We firmly believe that to answer the complex questions and concerns that emerge in daily practice, helping professionals need to understand, appreciate, and feel comfortable with the range of philosophic and methodologic traditions and the application of multiple design strategies to the health care setting.

CONTENTS

Part I

PHILOSOPHICAL FOUNDATON OF RESEARCH

Welcome to the world of research! For us, conducting research is one of the most challenging, creative, and intellectually satisfying professional activities. We believe that research is a professional responsibility. To develop and advance knowledge from which to base practice is essential if we, as health and human service professionals, are to provide services that enhance the quality of life of our clients and patients.

This text is divided into three major sections: Part I, "Philosophical Foundation of Research"; Part II, "Thinking Processes of Research Design"; and Part III, "Action Processes of Research." The sections move from an understanding of the philosophical foundations of the thinking and actions of researchers, to examining the language and thinking of researchers using diverse world views, to implementing a series of actions to obtain knowledge to answer your research questions.

We begin in Part I with our definition of research, a discussion of the forms of human reasoning, and an examination of the philosophical foundation of research designs and the types of designs that emerge from different world views. Knowing the philosophical base of the thinking and actions of researchers will enable you to understand and use the full spectrum of designs in your research.

We view research as a purposeful activity that can be accomplished in numerous ways. We have developed a conceptual framework to capture the multiple systematic strategies that health and human service professionals can use to generate or contribute to knowledge. We call this framework the "continua of research" and present a visual model of it in Chapter 2. The framework organizes our discussion of each thinking and action process that we present in this text. We have found it to be a useful and effective way of conceptualizing the world of research and have used it extensively in our own teaching and research. We hope this framework and book will be a valuable guide as you conceptualize and conduct research that serves your purposes and contributes to theory and practice in a thoughtful, creative way.

1 | Introduction: Research as a Way of Knowing

- **Why is Research Necessary?**
- **What is Research?**
- **Assumptions of Our Approach**
- **How to Use This Text**

WHY IS RESEARCH NECESSARY?

A 74-year-old woman with a fractured hip will shortly be discharged from rehabilitation, but appears unmotivated to use the self-care techniques you taught her. You wonder if rehabilitation has been effective and what her future functional capabilities will be on her return home.

You have just learned how to use a new tool to assess functional status in children. You wonder if this instrument is more accurate and useful than previous assessment instruments you have tried.

An interesting research article describes an effective discharge planning procedure for cognitively impaired adults. You wonder if you should implement these planning procedures in your own department.

You are initiating a new program in your professional practice to prevent low back injury in Hispanic migrant farm workers. To date, no existing prevention strategies have been effective in reducing the incidence of low back injury in that population. You want to know why traditional approaches have failed and how to develop an appropriate knowledge base from which to develop effective programs.

Helping professionals routinely have questions about their daily practice that are best answered through *systematic* investigation, or the research process. An important reason for conducting research is that it is a sound way to obtain scientific knowledge about the specific practice problems you may routinely experience.[1] A second important reason to participate in research is that this activity has become critical to the scientific advancement of your practice. In the health and human service arenas, the fundamental goal of research is to contribute to the development of a scientific body of knowledge of a profession. Research contributes to this goal in three basic ways. First, it generates relevant theory and knowledge about human experience and behavior; second, it develops and tests theories that form the basis of specific practices and treatment approaches, and third, it validates professional and health service delivery practices.[2]

There are other important reasons why you need to understand and participate in the research process. The knowledge we obtain through research is critical in guiding legislators and regulatory bodies about the best possible health and human service policies and services. Federal regulatory agencies and other fiscal intermediaries base many of their decisions and health and human service practice guidelines on empirical evidence or the knowledge generated through the research process.

It is also important to comprehend the research process to participate in research activities in your practice setting. In many health and human service settings, it is an expectation that professional departments participate in advancing the research goals of the institution. There are many diverse roles you may have as a member of a research team. You may initially want to participate in the process as a data collector, chart extractor, interviewer, provider of an experimental intervention, or recruiter of clients into a study. These are all excellent, time-limited roles to learn, first hand, the art and science of the research process. When you feel more comfortable and gain some experience with the process, you may want to serve as a project coordinator and be responsible for the coordination of the detailed tasks and daily activities of a research endeavor or as the co-investigator and assist in the conceptual development, design, implementation, and analytic components of a study. Finally, if you are really hooked on research, you may want to be a principal investigator and assume responsibility for initiating and overseeing the scientific integrity of the entire research effort.

It is becoming increasingly necessary for you to become a critical consumer of the growing body of research literature published in professional journals. Understanding research will provide you with the necessary skills to determine the adequacy of research outcomes and their implications for your daily practice.[3] The knowledge gained from research by you and others has the potential of improving the quality of life of the people that you serve—your patients or clients.

Finally, as helping professions engage in the research process, they not only contribute to the development of knowledge and theory and the validation of practice but also advance and refine the research process itself and its applications to professional issues. Health and human service professions involved in research today have been making significant contributions to the evolution of research methods.[4-7] Nevertheless, helping professionals are often hesitant to engage in research because of their unfamiliarity with and misconception of the process. Research is challenging, exhilarating, and very stimulating. However, just like many other professional activities, at times it can be time consuming, tedious, and frustrating. Both the challenge and the frustration emerge from research not being a simple activity, especially in health and human service domains. The complexity of both human behavior and the service environment, whether in the home, community, outpatient clinic, institution, or legislature, presents a different set of challenges for the researcher than other research settings and social research issues. Throughout this book, we will discuss the specific dilemmas and design implications posed by research conducted by helping professionals.

WHAT IS RESEARCH?

Research is not owned by any one profession or discipline. It is a systematic way of thinking and knowing and has a distinct vocabulary that can be learned and used by each helping profession.

There are many definitions of research, ranging from a very broad to very restrictive understanding of the research endeavor. A very broad conceptualization suggests that research includes any type of investigation that uncovers knowledge. On the other hand, a formal and more restrictive view, such as that offered by Kerlinger,[8] defines scientific research as "systematic, controlled, empirical, and critical investigation of natural phenomena guided by theory and hypotheses about the presumed relations among such phenomena [p. 10]. Whereas a broad definition includes any type of activity as research, Kerlinger's viewpoint implies that the only legitimate approach to scientific inquiry is that of hypothesis testing.

In contrast, in this text we define research as:

multiple, systematic strategies to generate knowledge about human behavior, human experience, and human environments in which the thought and action process of the researcher are clearly specified so that they are logical, understandable, confirmable, and useful.

This definition is unlike the one offered by the restrictive view in that we recognize the legitimacy and value of many distinct types of investigative strategies. It is also unlike the broad inclusive approach in that our definition clearly delineates the boundaries between research and other forms of knowing by establishing the four listed criteria: logical, understandable, confirmable, and useful. Let us examine the three major components of our definition.

Research as Multiple Systematic Strategies

The first component of our research definition recognizes the value of varied systematic strategies to understand the depth and range of research questions

TABLE 1–1.

Major Differences Between Naturalistic Inquiry and Experimental-Type Research

Domains	Naturalistic	Experimental Type
Epistemology	Multiple realities	Single objective reality
	Multiple epistemologies	Single epistemology
Primary thinking process	Inductive	Deductive
Purpose	Reveal complexity	Predict
	Uncover meanings of human experience	Explain
	Theory generating	Theory testing
Context	Natural setting	Controlled setting

asked by helping professionals. Our definition states that there is not one valid scientific methodology but many from which to examine complexity from multiple perspectives. These multiple research strategies have been categorized as either "naturalistic inquiry" or "experimental-type research." Each research strategy is founded in a philosophical tradition, follows a distinct form of human reasoning, and defines and obtains knowledge differently. In Table 1–1, we have summarized the major differences between these traditions in terms of their epistemology, or way of knowing. In Chapter 2 we explore these different philosophical traditions and the implications for research for the helping professional. We also suggest a third design category that combines both traditions, which is called "integrated designs."

Research as Thought and Action Processes

The second important component in our definition of research refers to thought and action processes. By thought and action processes we mean the different ways of reasoning and specific series of actions that distinguish naturalistic and experimental-type investigators in the conduct of their research. Experimental-type and naturalistic research strategies are founded in two distinct forms of human reasoning: *deduction* and *induction,* respectively.

Deductive Reasoning

Experimental-type researchers use deductive reasoning and begin with the acceptance of a general principle or belief and then apply that principle to explain a specific case or phenomenon. This approach in research involves "drawing out" or verifying what already is accepted as true.[9] For example, a researcher may start from the theory that caregiving involves ongoing stress and that health problems emerge when such stressors as inadequate knowledge of care strategies, years in caregiving, and few social supports are present. Accepting these principles as true, the deductive researcher is interested in testing the effectiveness of a series of interventions to reduce stress in caregivers as a means of improving their health status.

Inductive Reasoning

Researchers who work within a naturalistic framework primarily use inductive reasoning. This type of cognitive activity involves a process in which general rules evolve or develop from individual cases or observations of phenomena. Consider the same example of caregiving. The inductive researcher would be more interested in examining the life experiences of caregivers from their own perspective through observation and in-depth interviewing. From this approach, the researcher would develop a sense of what type of intervention would be most useful in promoting caregiver health. The researcher, proceeding inductively, seeks to reveal or uncover a truth based on the perceptions of caregivers. Intervention principles then would be developed based on these perceptions.

Difference in Knowledge

Although in any research project an investigator uses both types of reasoning, the overall research process can be characterized as following the structure of one or the other type of reasoning. Each type of reasoning may result in different knowledge. For example, the type of caregiver intervention that results from each form of reasoning may be totally different even though the focus, caregiving, is similar. Let us examine this point further.

The researcher working deductively assumes a truth before engaging in the research process and applies that truth to the investigation. In the caregiver example, the researcher assumes that all caregivers experience a form of stress and therefore would benefit from a stress-reduction intervention. The stress-reduction intervention is based on existing stress theory and knowledge and is planned and implemented by the researcher for all caregivers in the study. Interestingly, research that has tested individual and group psychotherapy sessions to reduce stress has reported that these type of interventions are mildly effective in reducing caregiver stress and only in some study participants.[10, 11] Although the intervention was somewhat helpful, the question remains as to why all caregivers did not benefit at the level at which the researcher had expected.

In a study proceeding inductively, the intervention emerges from what is learned from those who will be receiving it. Researchers following this inductive process have reported that caregivers express different levels of stress and identify other distinct issues and difficulties about their experience that cannot be explained by stress-theory or addressed by stress-oriented interventions.[12] Interventions that may be developed as a consequence of these insights may include a much broader array of services based on the specific needs and care issues identified by the individual.[11] If that is the case, the deductive researcher made an inaccurate assumption about the appropriateness and need of a stress-reduction intervention for all caregivers. The inductive researcher may have confirmed aspects of stress theory as it relates to caregivers but also demonstrated that the theory is too limited to understand the multiple needs of this group.

TABLE 1–2.

Major Characteristics of Inductive and Deductive Thinking

Inductive	Deductive
No a priori acceptance of truth	A priori acceptance of truth
Alternative conclusions can be drawn from data	One set of conclusions is accepted as true
Theory development	Theory testing
Examines relationships among unrelated pieces of data	Tests relationships among discrete phenomena
Development of concepts based on repetition of patterns	Testing concepts based on application to discrete phenomena
Holistic perspective	Atomistic perspective
Multiple realities	Single separate reality

An inductive reasoning approach in research is used to reveal theory, rules, and processes, whereas deductive reasoning is used to test or predict the application of theory and rules to specific areas of concern. Both approaches can be used to describe, explain, and predict phenomena, although traditionally only deductive reasoning has been valued as contributing to explanation and cause and effect relationships. Table 1–2 summarizes the major characteristics of each type of reasoning.

Research as Four Basic Characteristics

The third important component of our research definition refers to the criteria we use to characterize research. We have stated that scientific knowledge may be generated by multiple research strategies using inductive and deductive reasoning. However, any strategy must conform to four criteria of being logical, understandable, confirmable, and useful. Let us examine the meaning and significance of each criterion.

Logical

In research, there is a unique way of thinking and acting that distinguishes it from other ways in which we come to know about, understand, and make sense of our experiences. Charles Pierce,[13] one of the founders of the scientific research process, identified other ways of knowing as (1) authority, or being told by a respected or trusted source; (2) hearsay, secondhand information that is not verified; (3) trial and error, or knowledge gained through incremental doing, evaluating, and modifying one's actions to achieve a desired outcome; (4) history, knowing indirectly through collective past experiences; (5) belief, knowing without verification; (6) spiritual understanding, knowing through divine belief; and (7) intuition, explanations of human experience based on previous unique and personal organization of one's own experience.

The distinction between these other forms of knowing and research is that in the forms specified, it is not essential to systematically clarify the logical thinking and action processes by which the knowledge was developed; nor is it essential to base "knowing" on information that was derived systematically. Research, however, must be based on systematic thought processes, systematic investigative activities, findings, analysis, and conclusions.

By the word logical, we mean that the thought and action processes in research are clear and conform to accepted norms of deductive or inductive reasoning. Logic is a science that involves defined ways of thinking and relating ideas to develop an understanding of phenomena and their relationships. The systematic nature of research requires that the investigator proceed logically and articulate each thought and action throughout the research inquiry.

Understandable

It is not sufficient for a researcher to articulate a logical process. This process, the study outcomes, and conclusions need to make sense, be precise, be intelligible, and be credible to the reader or research consumer.

Confirmable

By confirmable we mean that the researcher clearly and logically identifies the strategies used in the study so that others can reasonably follow the path of analysis and arrive at similar outcomes and conclusions. The claims made by the researcher should be supported by the research strategy and be accurate and credible within the stated boundaries of the study.

Useful

Research generates, verifies, or tests theory and knowledge for use. In other words, the knowledge derived from a study should inform and, it is hoped, improve professional practice and client outcome. Each researcher, consumer, or professional judges the utility of a study based on his or her own needs and purposes. Usefulness is a subjective criterion in that it is based on one's judgment about the value of the knowledge produced by a study. However, the value of a study and the usefulness of knowledge become more widely accepted as the new knowledge increasingly stimulates further research and promotes the testing or verification of new or existing theory and practice.

ASSUMPTIONS OF OUR APPROACH

The definition of research that we use and the way in which this book is organized is based on the current thinking of researchers in diverse health and social science professions. Our approach significantly differs from previous texts in the following ways.

Most texts identify the scientific or experimental paradigm as the correct and valid methodologic approach from which to develop a professional knowledge base.[8, 14–17] Other texts focus on a single methodology within a naturalistic framework, such as ethnography[18–20] or grounded theory,[21] or present a discussion of different qualitative methods[22] and their use in research.[6, 7] Still other research texts for health and human service professionals explain naturalistic inquiry by using the framework or lens of the experimental researcher.[4, 23, 24] These latter texts tend to assume that the difference between naturalistic and experimental approaches exists only in respect to the specific procedures a researcher may follow.

Our viewpoint, however, reflects a new school of thought that has been expressed in numerous professional and academic disciplines. This school of thought proposes that both naturalistic and experimental-type research strategies are of equal importance to the establishment of a scientific base of health and human service practice and to the adequate examination of the diversity of human experience and behavior.[25–28] This school of thought also firmly asserts that it is not reasonable to critique naturalistic research using experimental language because each approach represents a distinct epistemology or way of knowing and obtaining knowledge.[22, 29, 30] As Gareth Morgan[31] eloquently claims:

> It is not possible to judge the validity or contribution of different research perspectives in terms of the ground assumptions of any one set of perspectives, since the process is self-justifying. Hence the attempts in much social science debate to judge the utility of different research strategies in terms of universal criteria based on the importance of generalizability, predictability and control, explanation of variance, meaningful understanding, or whatever are *inevitably flawed: These criteria inevitably favor research strategies consistent with the assumptions that generate such criteria as meaningful guidelines for the evaluation of research.* . . .Different research perspectives make different kinds of knowledge claims, and the criteria as to what counts as significant knowledge vary from one to another [p. 89].

In addition, this growing body of literature argues that it is not only possible but desirable and necessary to make explicit the specific and systematic methods used by qualitative researchers to advance these methodologies.[18, 22, 32, 33] As some have suggested, qualitative researchers need to develop standards of quality by which to judge the validity and effectiveness of naturalistic inquiry. How are we to distinguish, for example, a casual observation of emergency room behavior from a scholarly interpretation of underlying patterns?

In this book, we present the two research traditions of naturalistic inquiry and experimental-type research by using the philosophical framework and language of each. By uncovering the distinct and specific thinking and action processes of each framework, the reader will be able to adequately understand and accurately evaluate and critique a research study from within the epistemologic assumptions of the research itself. In this text, we suggest it is possible to explain the thought and action processes of naturalistic inquiry and to advance the standards by which such research is conducted, evaluated, and understood.

Although we may not solve the qualitative-quantitative controversy in this text, we do suggest that health and human service professionals need to transcend the debate and learn the purposes, value, and uses of each perspective. We also suggest that it is possible to carefully integrate approaches from these distinct philosophical traditions to answer select research questions.[33] In Chapter 11, we discuss this issue further and suggest that designs that draw from both the naturalistic and experimental-type schools of thought are developing as a paradigmatic category in their own right. This perspective represents a controversial new direction in health and human service research and has been referred to as "multimethod," "triangulated," or "integrated" designs. The reader will learn about the subtle differences among these terms and the ways in which some researchers are attempting to combine disparate philosophical traditions to answer specific research questions.

HOW TO USE THIS TEXT

Researchers face similar challenges and requirements, although these may be interpreted and acted on differently based on their philosophical viewpoint, form of human reasoning, or preferred research tradition. For the purposes of this text, we categorize these challenges and requirements into eight overlapping tasks:

1. Identifying a philosophical base and theoretical framework
2. Framing a research problem
3. Determining supporting knowledge
4. Selecting a design strategy
5. Implementing the research
6. Analyzing information
7. Reporting conclusions
8. Using the knowledge

Research Process

This text is organized along the eight challenges and basic requirements of research. We impose a linear sequence on our discussion, although these challenges are interrelated and may not necessarily occur in the order in which we present them. Each research task is discussed and understood from the viewpoint and language of both the researcher engaged in naturalistic inquiry and experimental-type research. Throughout our discussion, we draw on research examples from the health and human service professions and emphasize the meaning of a research principle for its application to the health and human service environment.

In Part I of this text we explore the initial research task, that of understanding the philosophical base of research and level of theory development in one's area of interest. This is perhaps the most misunderstood and feared aspect of research. Although it is not our intent to explore the history of philosophy and the current debates on the philosophy of science, we do believe that a basic understanding of these traditions and the relationship between theory and research is critical and will make you a more thoughtful and skilled researcher. These concepts are explored in Chapters 2 and 3.

In Part II of this text we explore the basic thinking processes in research. First, the researcher needs to identify a topic and frame a research problem that can be submitted to systematic investigation. This process of question formation is explored in Chapter 4, where a distinction is made between the broad research query initially posed and the framing and reframing process that occurs in naturalistic inquiry and the specific, concise research question developed for experimental-type research.

Determining supportive knowledge (or conducting a literature search) for your particular research problem usually occurs in tandem with developing and refining your research question or query. We discuss this task in Chapter 5 and provide specific suggestions as to how to logically organize a search. Selecting a design strategy is perhaps the most fundamental aspect of the research process. Based on one's philosophical position, research purpose, specific query, and supporting literature, the researcher develops a suitable set of methods to accurately respond to the research query or question. In Chapter 6, we discuss the basic concept of design and how it is used differently in naturalistic inquiry and experimental-type research. Although it would be impossible to discuss in detail the multiple design strategies available, Chapters 7 and 8 present basic experimental-type designs, and Chapters 9 and 10 discuss basic designs that characterize naturalistic inquiry. In Chapter 11, we present the concept of ''integrated designs'' and the way in which different methods and designs can be combined to address a single research focus. We address integrated design in one chapter rather than throughout the text for several reasons. First, an understanding of the thought and action processes of experimental-type and naturalistic designs is essential before one can consider the complexities of integration. Second, integrated designs build on experimental-type and naturalistic strategies. Including integrated approaches in each chapter would therefore be somewhat redundant. We hope the organization of this text will stimulate your thinking as to how strategies from both methodological schools of thought can be used in tandem to answer your research questions.

In Part III we explore the action processes of implementing research, the fifth research challenge. In Chapter 12, we discuss how the researcher sets boundaries to limit the scope of the inquiry so that it is "doable." Boundary setting strategies are discussed for both experimental-type and naturalistic studies. Then in Chapters 13 and 15 we examine approaches by which to obtain information. The researcher may choose from a variety of techniques that range along a continuum from unstructured looking and listening to structured, fixed choice asking in the form of questionnaires.

Analysis of information, the sixth research challenge, involves a series of planned activities that again differ depending on the type of tradition or paradigm from which one is working. These differences are explored in Chapter 16. Chapter 17 introduces you to the range of statistical techniques for quantitative information, and Chapter 18 examines analytic techniques for qualitative and narrative information.

Reporting and using research are necessary tasks that complete the research process. Reporting conclusions involves the process of disseminating the research process and results. As discussed in Chapter 19, it may take several different forms, depending on the nature of the research pursued. Using knowledge generated by the research involves understanding how to implement research findings and the relationship between research outcomes and professional practice. We conclude the book by sharing with you some of our more memorable experiences in conducting varied projects in Chapter 20.

Within these eight basic research challenges, health and human service researchers face ethical dilemmas and research application issues. Ethical concerns focus on the rights of human subjects, the conduct of the investigator, and the ethics of the question and design procedures. These issues are explored throughout the book.

Who Should Read This Text?

This book can be used by undergraduate and advanced students in the health and human service professions as a basic introduction to the steps of the research process pursued by researchers who are engaged in both naturalistic and experimental-type research. It can also be used by practitioners and beginning researchers who would like to broaden their understanding of the range of methodologies available to answer their questions. The text is adaptable to both the classroom and practice setting and can assist helping professionals to develop strategies to pursue their research ideas. Innovative features of this book include:

- Discussion of each aspect of the research process from idea identification to use of computers and manuscript development from the perspective of the experimental and naturalistic researcher;

- Application of each step of the research process to the practice environment;
- A bibliography of articles and books from which to pursue in-depth reading of each major point raised in this text.

Conducting research in the practice environment and answering practice-driven research questions present unique challenges. We hope this book is instructive and that it provides you with the basic logic of different research traditions to examine and solve the health and human service issues that develop in your practice.

REFERENCES

1. Yerxa E: Seeing a relevant, ethical and realistic way to knowing for occupational therapy, *Am J Occup Ther* 45:199, 1991.
2. Payton OD: *Research: the validation of clinical practice,* ed 2, Philadelphia, 1988, FA Davis.
3. Ottenbacher KJ: *Evaluating clinical change: strategies for occupational and physical therapists,* Baltimore, 1986, Williams & Wilkins.
4. Burns N, Grove S: *The practice of nursing research: conduct, critique and utilization,* Philadelphia, 1987, WB Saunders.
5. Whyte WF: *Participatory action research,* Newbury Park, Calif, 1991, Sage.
6. Parse RR, Coyne AB, Smith MJ: Nursing research traditions: quantitative and qualitative approaches. In *Nursing research: qualitative methods,* Bowie, Md, 1985, Brady Communications, pp 1–8.
7. Morse J: *Qualitative nursing research,* Rockville, Md, 1989, Aspen.
8. Kerlinger FN: *Foundations of behavioral research,* ed 2, New York, 1973, Holt, Rinehart, & Winston.
9. Hoover KR: *The elements of social scientific thinking,* ed 2, New York, 1980, St Martin's Press.
10. Zarit SH: Do we need another caregiver study? *Gerontologist* 29:47, 1989.
11. Corcoran MA, Gitlin LN: Environmental influences on behavior of the elderly with dementia: principles for home intervention, *Occup Ther Geriatr* 4:5–22, 1991.
12. Hasselkus B: Meaning in family caregiving: perspectives on caregiver/professional relationships, *Gerontologist* 28:686, 1988.
13. Peirce CS: *Essays in the philosophy of science,* Indianapolis, 1957, Bobbs-Merrill.
14. Currier DP: *Elements of research in physical therapy,* Baltimore, 1984, Williams & Wilkins.
15. Oyster CK, Hanten WP, Llorens LA: *Introduction to research: a guide for the health science professional,* Philadelphia, 1987, JB Lippincott.
16. Babbie E: *The practice of social research,* ed 6, Belmont, Calif, 1992, Wadsworth.
17. Cox RC, West WL: *Fundamentals of research for health professionals,* Laurel, Md, 1982, Ramsco.
18. Agar MH: *Speaking of ethnography,* Newbury Park, Calif, 1986, Sage.
19. Spradley JP: *Participant observation,* New York, 1980, Holt, Rinehart, & Winston.

20. Geertz C: *The interpretation of cultures,* New York, 1973, Basic Books.
21. Glaser BG, Strauss AL: *The discovery of grounded theory: strategies for qualitative research,* New York, 1967, Aldine.
22. Lincoln YS, Guba EG: But is it rigorous? Trustworthiness and authenticity in naturalistic evaluation. In Williams DD, editor: *Naturalistic evaluation,* San Francisco, 1986, Jossey-Bass.
23. Stein F: *Anatomy of clinical research: an introduction to scientific inquiry in medicine, rehabilitation and related health professions,* Thorofare, NJ, 1989, Slack.
24. Bailey D: *Research for the health professional: a practical guide,* Philadelphia, 1991, FA Davis.
25. Evolving methodology in disability, *Rehabil Briefs* XII:5, 1989.
26. *American Evaluation Association: Health Evaluation Newsletter,* 1990, pp 1–2.
27. Pearlin L: Structure and meaning in medical sociology, *J Health Soc Behav* 33:1–9, 1992.
28. Zemke R: The continua of scientific research designs, *Am J Occup Ther* 43:551–553, 1989.
29. Smith JK: Quantitative vs. qualitative research: an attempt to clarify the issue, *Educ Researcher* 112:6–13, 1983.
30. Guba EG: Criteria for assessing the trustworthiness of naturalistic inquiries, *Educ Commun Technol J* 29:75–91, 1981.
31. Patton M: *Qualitative evaluation and research methods,* ed 2, Newbury Park, Calif, 1990, Sage.
32. Salomon G: Transcending the qualitative-quantitative debate: the analytic and systematic approaches to educational research, *Educ Researcher* 20:10–18, 1991.
33. Miles MB, Huberman AM: *Qualitative data analysis,* Newbury Park, Calif, 1984, Sage.

2

Continua of Research

- **Philosophical Base of Experimental-Type Research**
- **Philosophical Base of Naturalistic Inquiry**
- **Implications of Philosophical Differences for Design**
- **Continua of Designs**
- **Selecting a Design Strategy**
- **Summary**

In Chapter 1 we introduced you to the way in which we define research and the differences between deductive and inductive reasoning. In this chapter we expand this discussion and explore the philosophical foundations of experimental-type research and naturalistic inquiry.

Although you do not have to be a philosopher to engage in the research process, it is important to understand the philosophical foundations and assumptions about human experience and knowledge of experimental-type and naturalistic research. By being aware of these assumptions, you will be more skillful in directing the research process and selecting specific methods to use and combine. Also, an understanding of these philosophical foundations will help you see that "knowledge" is determined by the way you frame a research problem and the strategy you use to obtain information.

The questions "What is reality?" and "How do we come to know it?" have been posed by philosophers and scholars from many academic and professional disciplines throughout history. As we have already suggested, there are two competing views of reality and how to obtain knowledge that reflect the basic differences between naturalistic inquiry and experimental-type research. Logical positivism is the foundation for deductive, predicting designs we refer to in this text as experimental-type research. On the other hand, a number of holistic and humanistic philosophical perspectives use inductive reasoning and are referred to throughout this text as naturalistic inquiry. Let us examine these philosophical traditions.

PHILOSOPHICAL BASE OF EXPERIMENTAL-TYPE RESEARCH

Experimental-type researchers share a common frame of reference or epistemology that has been called rationalistic, positivist, or logical positivism. Although there are theoretical differences between these terms, we use the term logical positivism to name the overall perspective on which deductive research design is based.

David Hume, an 18th century philosopher,[1] was most influential in developing this traditional theory of science. This viewpoint posits that there is a separation between individual thoughts and what is real in the universe outside ourselves. That is, traditional theorists of science define knowledge as part of a reality that is separate and independent from individuals and verifiable through the scientific method. They believe the world is objectively knowable and can be discovered through observation and measurement that is considered "unbiased."[2] This epistemologic view is based on a fundamental assumption that it is possible to know and understand phenomena that reside outside of ourselves, separate from the realm of our subjective ideas. Only through observation and sense data, defined as that information obtained through our senses, can we come to know truth and reality.

Philosophers in subsequent centuries further developed, modified, and clarified Hume's basic notion of "empiricism" to yield what today is known as logical positivism. Essentially logical positivists believe that there is a single reality that can be discovered by reducing it into its parts. The relationship among these parts and the logical, structural principles that guide them can also be discovered and known through the collection and analysis of sense data, leading finally to the ability to predict phenomena from that which is already known.[2] Bertrand Russell,[3] a 20th century mathematician and philosopher, was then instrumental in promoting the synthesis of mathematical logic with sense data. Statisticians such as Fischer and Pearson developed theories to logically and objectively reveal "fact" through mathematical analysis. The logical positive school of thought therefore provided the foundation for what most lay persons have come to know as experimental research. In this approach, a theory or set of principles is held as true. Specific areas of inquiry are defined and hypotheses, or expected outcomes of an inquiry, are posed and tested. Sense data are then collected and mathematically analyzed to support or refute hypotheses. Through incremental deductive reasoning, which involves theory verification and testing, "reality" can become predictable. Another major tenet of logical positivism is that objective inquiry and analysis are possible. That is to say, the investigator, through the use of accepted and standard research techniques, can eliminate bias and achieve results through objective, quantitative measurement.[4]

PHILOSOPHICAL BASE OF NATURALISTIC INQUIRY

Another school of theorists has argued the opposite position: that individuals create their own subjective realities and thus knower and knowledge are interrelated and interdependent.[2, 5, 6] This group believes ideas are the lens through which each individual knows the universe and that it is through these ideas that we come to understand and define the world. This epistemologic viewpoint is based on the fundamental assumption that it is not possible to separate the outside world from an individual's ideas and perceptions of that world. Knowledge is based then in how the individual perceives experiences and understands his or her world. A number of research strategies share this basic holistic epistemologic view, although each is rooted in a different philosophical tradition.

Although research based on holistic perspectives is relatively new, these perspectives are not. Ancient Greek philosophers struggled with the separation of idea and object, and philosophers throughout history continued that debate.[7] The essential characteristics of what we are calling holistic philosophies are as follows: human experience is complex and cannot be understood by examining only its parts, meaning in human experience derives from an understanding of individuals in their social environments, multiple realities exist and one's view of reality is determined by events viewed through individual lenses or biases, and those who have the experiences are the most knowledgeable about them.

In addition to these common characteristics, holistic philosophies encompass a number of principles that guide the selection of particular designs in this category. For example, phenomenologists believe that human meaning can be understood only through experience.[8] Thus, a phenomenologic understanding is limited to knowing experience without interpreting that experience. On the other hand, interpretive interactionists[9] assume that human meaning evolves from the context of social interaction. Human phenomena can therefore be understood through interpreting the meanings in social discourse and exchange. The holistic philosophies therefore suggest a pluralistic view of knowledge or that there are multiple realities that can be identified and understood only within the natural context in which human experience and behavior occur. Coming to know these realities requires research design that investigates phenomena in their natural contexts and seeks to discover complexity and meaning. Furthermore, those who experience are the "knowers" and transmit their knowledge through doing and telling.[6]

IMPLICATIONS OF PHILOSOPHICAL DIFFERENCES FOR DESIGN

How you implicitly define knowledge and the relationship between the knower (researcher) and the known (research outcome or phenomena of your study) direct your entire research effort: from framing your research question to reporting your findings. Let us consider some examples of these important philosophical concepts and their implications for research design.

Let us suppose two researchers want to know about what happens to participants in group therapy who have joined the group to improve their self-confidence. The researcher who suggests that knowing can be objective might choose a strategy in which he or she defines self-confidence as a score on a pre-existing scale and then would measure participants' scores at specified time intervals to ascertain changes. Changes in the scores would tell the investigator what changes the group experienced on that scale.

The investigator who believes that the world can be known only subjectively might choose a research strategy in which the group is observed and the members interviewed to obtain their perspective on their own process within the group. A third researcher, who believes in the value of many ways of knowing, may integrate strategies to understand the nature of the change from the perspective of the participants and from the perspective of existing theory as measured by a self-confidence scale.

CONTINUA OF DESIGNS

We have indicated that there are two primary philosophical traditions in research, which can be categorized as either experimental type or naturalistic. Within each of these categories we have said that there are many systematic research strategies. We would now like to suggest that research strategies fall along two primary design continua. First is the continuum of strategies that share the philosophical foundation of experimental-type research in structure, degree of control, and purpose which can be ordered according to these principles. Throughout this text we refer to this as the continuum of experimental-type research.

On the other hand, strategies that share the principles of a holistic perspective or naturalistic inquiry can be ordered along a separate continuum that reflects differences in purpose, degree of investigator involvement, and investigator-imposed structure and control. Throughout this text, we refer to this as the continuum of naturalistic inquiry. Each design continuum has its own language, thought and action processes, and specific design issues and concerns. We use the concept of the two continua throughout this text as a basis from which to discuss and compare design structures and research processes within the same philosophical perspective and across philosophical traditions.

- Extent of investigator
  imposed structure
- Degree of control
- Degree of manipulation

- Experimental
- Quasi-experimental
- Pre-experimental
- Non-experimental
- Descriptive
- Exploratory

FIG 2–1.
Continuum of experimental-type design.

Continuum of Experimental-Type Research

Many designs follow the philosophical tradition of logical positivism. Figure 2–1 displays the six major categories of designs that form this continuum: (1) exploratory, (2) descriptive, (3) nonexperimental, (4) pre-experimental, (5) quasi-experimental, and (6) experimental method.[10–12] These designs range along the continuum according to their purpose and degree of control, manipulation, and investigator-imposed structure. At the bottom of the continuum are exploratory and descriptive designs. Designs of this type have the primary purpose of exploration and description and have the least degree of control, manipulation, and investigator-imposed structure. In the middle of the continuum are nonexperimental-type designs that seek an understanding of relationships and offer some degree of investigator-imposed control and structure. At the top of the continuum are experimental methods that have the primary purpose of predicting and identifying causal relationships. These designs have the most degree of control, manipulation, and investigator-imposed structure. Studies that use any one of these design structures share the same assumptions about reality and use deductive reasoning. Each design type is discussed in Chapters 7 and 8.

Continuum of Naturalistic Inquiry

There is great diversity in the strategies that are categorized as naturalistic inquiry. As we indicated earlier in this chapter, the designs that form this continuum are each rooted in a different philosophical tradition and theoretical perspective. The language and thought processes used by researchers along this continuum vary within each design structure, as well as being quite different from the language and thought processes of researchers along the experimental-type continuum. However, they all have characteristics in common and use qualitative methodologies, although often for different purposes, and to answer very distinct types of questions. Figure 2–2 displays the major categories of design that form this continuum: (1) endogenous, (2) critical theory, (3) phenomenology, (4) heuristic design, (5) life history, (6) ethnography, and (7) grounded theory.

These designs range along the continuum according to the extent to which inquiry involves the personal "essence," experience, and insights of the investigator, the extent to which individual "experience" vs. patterns of human experience is sought, and the extent to which the investigator imposes structure in the data collection and analytic processes. At the bottom of the continuum are designs such as endogenous and critical theory that involve the investigator turning over so to speak the control and direction of exploration to those individuals who are being studied. In the middle of the continuum are designs such as phenomenology, heuristic, and life history, in which the investigator's own experiences and collaborative relationship with informants structures the nature of the inquiry. At the top of the continuum are designs such as ethnography and grounded theory in which the investigator seeks to uncover patterns of behavior and human experience and imposes more direction and structure on the data collection and analytic process.

This approach in categorizing and arranging designs along a continuum is somewhat arbitrary and reflects the way we prefer to understand and compare designs using naturalistic inquiry. For example, we have placed endogenous research at the bottom of the continuum because its basic intent is to turn the research process over to those who are also being investigated. However, as you will learn in Chapter 10, a variety of data collection approaches may be selected in endogenous research, including those that are highly structured and impose

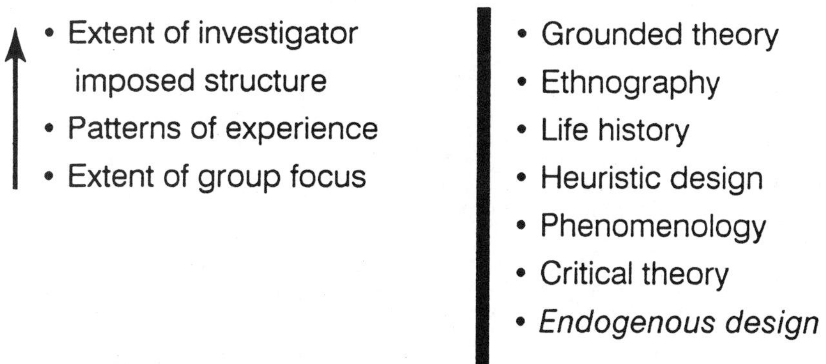

- Extent of investigator imposed structure
- Patterns of experience
- Extent of group focus

- Grounded theory
- Ethnography
- Life history
- Heuristic design
- Phenomenology
- Critical theory
- *Endogenous design*

FIG 2–2.
Continuum of naturalistic inquiry.

control by those conducting the investigation. Therefore, some might argue that endogenous research could be placed at the bottom or top of the continuum according to the specific strategies that are developed.

Integrated Designs

It is possible to use a variety of strategies in a research study. Strategies used in one design can be combined with those used in another, and these designs may be either along the same continuum or across continua. There are many reasons to combine or integrate designs, and these are discussed in Chapter 11. To understand integrated strategies, we suggest that they too can be categorized along a continuum of their own. This continuum ranges according to the degree to which strategies are integrated throughout the research process. Figure 2–3 displays the major categories of strategies: (1) data collection efforts, or triangulation; (2) mixed method strategies; and (3) fully integrated strategies.

At the bottom of the continuum are data collection efforts that use triangulation. This strategy represents the least degree of integration in a design. In this approach, different ways of collecting information are combined to obtain insight on the same phenomenon. Usually triangulation is used to strengthen the credibility of one's findings. As we move up the continuum, the degree of integration in a design increases in mixed method approaches. In a design using mixed methods, the researcher uses one framework—either that of experimental or of naturalistic inquiry. Within that framework, action processes are borrowed from either the naturalistic or experimental-type continua to answer a single research question or query. For example, in a study of how women who are intravenous drug users cope with human immunodeficiency virus, Suffet and Lifshitz,[13] used a naturalistic framework. However, they combined experimental-type sampling techniques to strengthen their study.

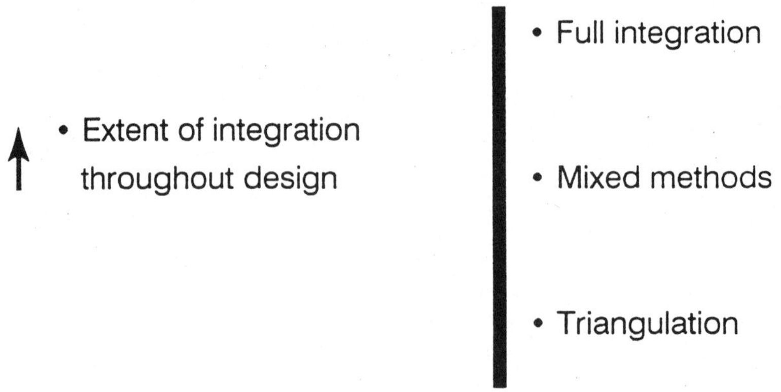

- Full integration

↑ • Extent of integration
   throughout design

- Mixed methods

- Triangulation

FIG 2–3.
Continuum of integrated strategies.

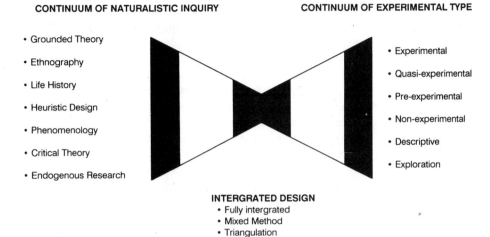

CONTINUUM OF NATURALISTIC INQUIRY   CONTINUUM OF EXPERIMENTAL TYPE

- Grounded Theory
- Ethnography
- Life History
- Heuristic Design
- Phenomenology
- Critical Theory
- Endogenous Research

- Experimental
- Quasi-experimental
- Pre-experimental
- Non-experimental
- Descriptive
- Exploration

INTERGRATED DESIGN
- Fully intergrated
- Mixed Method
- Triangulation

FIG 2–4.
Continua of research design.

At the top of the continuum are fully integrated designs. These are intended to both verify and develop new theory. Such integrated designs use the frameworks of distinct philosophical traditions to answer different questions within one study.[14] For example, this type of integration allows an investigator to test an intervention based on well-developed theory using experimental-type design. It also permits discovery of new insights as to the underlying process of change as a consequence of the intervention using a naturalistic design. The language and thinking and action processes on the integrated design continuum are diverse in that these designs draw from the full spectrum of research strategies available and are only limited by the creativity of the researcher.

As you read about the two primary continua, experimental type and naturalistic, throughout this book, we encourage you to think creatively about combining strategies and about how you may use varied strategies to address research problems.

In Figure 2–4, we show you the complete model that guides this text. Each chapter or subsection within a chapter focuses on a part of the continua of research design. This focus is highlighted in black on the graphic throughout the text, and can guide you in your reading. We have placed integrated designs in the middle of the graphic with connections to both the naturalistic and experimental-type continua. We view these lines as fluid and suggest that the placement of integrated designs could vary and be placed either closer to the left (naturalistic continuum) or to the right (experimental-type continuum) of the graphic, depending on the specific strategies that are used and combined in a study.

There are many other ways of categorizing and arranging designs. In Chapter 6, we share with you how others have done so for a point of comparison.

We also encourage you to think of different categorical schemes for both experimental-type and naturalistic inquiry for the purposes of understanding and appreciating the unique qualities of each design.

The model of the continua of research in Figure 2–4 serves as a fundamental organizing perspective from which to understand and appreciate the thinking and action processes of researchers working from very different world views. The placement of a design at either the top, middle, or bottom along the experimental-type or naturalistic continuum does not indicate the adequacy or value of a design. As we shall repeatedly remind you throughout this book, there is nothing inherently good or bad about any one design type; rather, each has value. The merits of a design lies in its ability to answer your specific research query or question in light of the realities of the field in which the study is implemented.

SELECTING A DESIGN STRATEGY

How does one decide which continuum to work from and which particular design strategy to choose? The selection of a design strategy is based on three considerations: (1) what one wants to do or one's purpose in conducting the research, (2) the way in which one thinks or reasons about phenomena, and (3) the level of knowledge development in the area to be investigated.

Purpose of Research

Research is a purposeful activity; that is, research is conducted for a specific reason, to answer a specific question, to solve a particular controversy or issue. We have not yet met anyone who wakes up one morning and says, "I just want to do a research project." Purpose drives the decision to engage in research. For example, you might need to conduct a research project to validate a health care intervention to obtain necessary resources and continued support for the program. Or perhaps you want to conduct a research study to help you systematically determine the elements of a new therapeutic approach in a previously untested area. Let us consider another example of how purpose drives the selection of a design.

Consider a case in which an administrator of a rehabilitation unit in a large teaching and research hospital has just been informed that her social work unit may be closed if she cannot demonstrate the value of the department's programs. The administrator decides that she should implement a research project to demonstrate the value of the service to the patients who are being discharged. She selects three criteria for measurement to define valuable intervention: (1) patient satisfaction, (2) social work staff time spent with patient's family in preparation for discharge, and (3) daily living skill performance level in the occupational therapy clinic before discharge. She selects and measures these three criteria on an a priori basis or deductively because she knows that these are the strengths of her unit staff that can be clearly demonstrated. Her purpose in conducting the research is to enhance the chances of survival of her unit. She selects a deductive strategy based on her previous, a priori knowledge of the strengths of her programs, and the types of positive outcomes she hopes to achieve.

In another hospital, the rehabilitation administrator has been asked to develop new programming for persons with acquired immunodeficiency syndrome (AIDS) and their families. In the absence of a well-developed body of knowledge in this area, the administrator selects an inductive strategy to find out what type of rehabilitation programming would be most valuable. She conducts in-depth, unstructured interviews with persons with AIDS, their families, and service providers to reveal needs and the methods by which those needs can be addressed within her institution.

As you can see, purpose is a powerful force that drives the selection of a research strategy. In the first example, the administrator needed to select a strategy that she as the researcher could control and that was based on the a priori assumptions she was willing to make. In the second example, the administrator needed to select a strategy that would uncover information that had not been determined previously. Because patient needs as perceived by patients and family were shaping institutional programming, an inductive, naturalistic design strategy was selected.

It is not uncommon to have more than one purpose for your study. Perhaps you already know that your intervention is successful, but you want to improve a service or obtain an understanding as to why the intervention is effective. This dual purpose may lead to a study that integrates a design from the experimental-type continuum with a design from the naturalistic inquiry continuum. The purpose of such an integrated study would be to verify what is valuable, as well as to reveal new avenues for programmatic improvement and explanation of effectiveness.

Preference for Knowing

One's preferred way of knowing is a second important consideration in selecting a particular category of research design. Selecting a design strategy is based, in part, on one's implicit view of "reality" (ontology), one's view of how we know what we know (epistemology), and one's comfort level with epistemologic assumptions. Our preferred way of knowing is often not something on which we consciously reflect. It is usually demonstrated in what approach to research makes the most sense to us and feels more comfortable. Investigator personalities also often influence the inclination to participate in one of the three continua.

Level of Knowledge Development

The third consideration involves the level of knowledge that has been developed in the particular topic area of interest. When little to nothing is understood about a phenomenon, a more descriptive and naturalistic approach to inquiry is most appropriate. When a very well-developed body of knowledge has been advanced, predicting designs may be more appropriate and suit the purposes of the investigator. However, this is not to suggest that naturalistic inquiry is chosen only when little knowledge has been developed. Again, choice is based on the combination of level of knowledge and the investigator's purpose. We explore the relationship of knowledge to design selection in more detail in the next chapter.

SUMMARY

We have made five major points in this chapter:

- Experimental-type research is based in a single epistemological framework of logical positivism and involves a deductive process of human reasoning.
- Naturalistic inquiry is based in multiple philosophical traditions that can be categorized as holistic in their perspectives and involves an inductive process of human reasoning.
- Integrated research draws on strategies from both experimental-type and naturalistic continua and may involve multiple and varied thought and action processes.
- Research strategies range along two primary continua: experimental-type or naturalistic inquiry. Those continua can be used in tandem with one another to form a third continuum of integrated design. Each category along the continuum is anchored in purpose, existing relevant theory, and a philosophical view of knowledge.
- The selection of a strategy from either continuum is based on one's preferred way of knowing, one's research purpose, and the level of knowledge development in the topic area of interest.

REFERENCES

1. Hume D: In Bigg LAS, editor: *A treatise on human nature,* New York, 1902, Clarendon Press.
2. Guba EC: Carrying on the dialogue. In Guba EC, editor: *The paradigm dialogue,* Newbury Park, Calif, 1990, Sage.
3. Russell B: *Introduction to mathematical philosophy,* New York, 1919, Macmillan.

4. Hoover KR: *The elements of social scientific thinking,* ed 2, New York, 1980, St Martin's Press.
5. Reason P, Rowan J: *Human inquiry,* New York, 1981, John Wiley & Sons.
6. Orcutt BA: *Science and inquiry in social work practice,* New York, 1990, Columbia University Press.
7. Randall JH, Buchler J: *Philosophy: an introduction,* New York, 1962, Barnes & Noble.
8. Husserl E [Boyce WR, translator]: *Ideas: general introduction to pure phenomenology,* New York, 1931, Macmillan.
9. Denzin N: *Interpretive interactionism,* Newbury Park, Calif, 1988, Sage.
10. Babbie E: *The practice of social research,* Belmont, Calif, 1992, Wadsworth.
11. Cook TD, Campbell DT: *Quasi-experimentation: design and analysis issues for field settings,* Boston, 1979, Houghton Mifflin.
12. Campbell DT, Stanley JC: *Experimental and quasi-experimental designs for research,* Boston, 1963, Houghton Mifflin.
13. Suffet F, Lifshitz M: Women addicts and the threat of AIDS, *Qualitative Health Res* 1, 51–79.
14. DePoy E, Gallagher C, Calhoun L, et al: Altruistic activity vs. self-focused activity: a pilot study, *Topics Geriatr Rehabil* 4:23–30, 1989.

3

Relationship Between Theory and Research

- **Why Is Theory Important?**
- **What Is Theory?**
- **Theory in Experimental-Type Research**
- **Theory in Naturalistic Inquiry**
- **Summary**

The level of knowledge and theory development in a particular field or in your area of interest determines, in part, the type of design you choose and the specific methods and analyses of a study. In this chapter we discuss the relationship of theory and research and explain the way in which theory is used by both the experimental-type and naturalistic researcher.

WHY IS THEORY IMPORTANT?

When people hear the word "theory," they often feel overwhelmed and assume that the term is not relevant to their daily lives or is too abstract and complicated to understand, but think how difficult life would be without theory. Theory informs us each day in many aspects of our existence, from knowing how to prepare for the weather to guiding our professional practice. For example, predicting the weather is based on existing theory. The meterologist looks at present weather conditions and past weather patterns and makes a prediction in the form of a forecast. In health and human service delivery, you make a decision about which intervention to use based on theory. You use a theory to guide your decisions even though you may not be fully cognizant that you are actually doing so. Let us suppose you are a social worker in a psychiatric hospital. A new member comes to your group therapy session with a diagnosis of major depressive disorder. Theoretically you know that this diagnosis is associated with a set of symptoms and behaviors, particularly self-degradation and melancholia. Based on this theoretical understanding of the underlying symptoms and the prediction of the individual's behavioral outcomes, you approach this client carefully and supportively. Without the theory, you would not know how to skillfully work with this client.

We have said in Chapter 1 that a primary purpose of research is to test theory using deduction and experimental-type designs, or to generate theory using inductive, naturalistic-based designs. When theory is well developed and fits the phenomenon under investigation, deductive studies are appropriate. When theory is poorly developed or does not fit the phenomenon under study, inductive studies that generate theory are more appropriate. The degree of knowledge development and relevance of a theory for the particular phenomenon under investigation determines in part the nature and type of research that you will do. Let us consider an example that reflects the differences in knowledge development and use of research to test and build theory.

Lawrence Kohlberg[1] is a well-known scholar who developed a theory of moral reasoning. He suggested three basic levels of moral reasoning that develop and unfold throughout the life span: preconventional, conventional, and postconventional. Briefly, in the preconventional reasoning stage, individuals develop notions of right and wrong based on reward or punishment; in conventional reasoning, individuals use accepted norms of right and wrong to guide their moral decisions; and in postconventional reasoning, humans use their own intrinsic notions of morality to formulate decisions and actions. Many investigators have accepted this theory as truth and have attempted to characterize and predict the moral reasoning of different populations. Scales to measure moral reasoning were developed based on Kohlberg's theory and used with different groups of people representing diverse socioeconomic, age, and gender groups. Based on these studies, populations and individuals were categorized regarding the level of moral reasoning that they exhibited. This information was extremely valuable in promoting our understanding of morality in humans. The characterization of moral reasoning based on Kohlberg's theory is an example of a deductive relationship between theory and research. That is to say, the theory has been accepted as true, and subsequent studies have intended to verify and advance the application of the theory in varied populations.

However, in the majority of studies in adults, women were found to reason at a lower level than men. Researchers concluded that women were not as morally advanced as men. Carol Gilligan,[2] a feminist researcher, did not accept Kohlberg's theory as true for women because it was intuitively contrary to what she experienced and knew. Even though knowledge was well developed in this area, the theory did not seem to fit or be relevant to the experiences of women. She used inductive research strategies to discover and understand the nature of moral reasoning in women. Through extensive in-depth interviews with women, Gilligan found that women were not less moral than men but used different criteria on which to base their moral decision making. These criteria were substantially different from those identified by Kohlberg's theory and those operationalized in the scales that had been developed to reflect his theory. Thus, although there was extensive theory development, Gilligan's research suggested that Kohlberg's notions did not explain moral reasoning in women. Inductive strategies were essential to reveal the different processes by which women reasoned and to develop theory that reflected their thought processes. Not surprisingly, Kohlberg had developed his theory of moral reasoning by studying men. Therefore, the deductive strategy was most valuable in expanding knowledge of the moral reasoning in men but not necessarily in women. Gilligan's research strategy illustrates the relationship between theory and naturalistic inquiry. In other words, naturalistic inquiry is extremely valuable as a theory-generating tool, but it also can be used to verify or refute a theory.

WHAT IS THEORY?

So what is a theory? Numerous definitions of theory can be found in research texts and books on theory construction. We believe Kerlinger's definition of theory is the most comprehensive and useful. Kerlinger[3] defines theory as "a set of interrelated constructs, definitions and propositions that present a systematic view of phenomena by specifying relations among variables, with the purpose of explaining or predicting phenomena" [p. 9]. In this definition, theory is a set of related ideas that have the potential to explain or predict human experience in an orderly fashion and that are based on data. The theorist develops a structural map of what is observed and experienced in an effort to promote understanding and to facilitate the ability to predict outcomes under specific conditions. Through research or empirical investigations, theories are either supported and verified or refuted and falsified.

As implied in Kerlinger's definition, there are four interrelated components subsumed under theory that range in degree or level of abstraction. Let us examine each level.

Levels of Abstraction

The word "abstraction" often conjures up a vision of the ethereal, the "not real." In this book, however, we use the term abstraction as it relates to theory development, to depict symbolic representation of shared experience. For example, if we all see a form that has fur, has a tail, has four legs, and barks, we name that observation a dog. We have shared in the visual experience and have created a symbol (the word dog) to name that sensory experience. All words are merely symbols to describe shared experience. Words are only one form of abstraction and different words represent different levels of abstraction.

Figure 3–1 displays four levels of abstraction within theory and the relationship of each to the other. Shared experience is the foundation on which

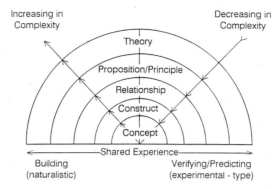

FIG 3-1.
Levels of abstraction.

abstraction is built. It is important to recognize that we do not use the word "reality" as the foundation. That term would imply that there is only one viewpoint from which to build the basic elements of a theory. Rather, this book is based on the premise that humans experience multiple realities and multiple perspectives about the nature of reality. As such, levels of abstraction must be built on shared experience, defined as the consensus of what we obtain through our senses. For experimental-type researchers working deductively, shared experience is usually thought of as sense data or that which can be reduced to observation and measurement. For researchers working inductively within the naturalistic paradigm, shared experience may include meanings and interpretations of human experience. Let us use the example of the furry being with four legs and a tail who barks to illustrate. Shared experience tells each of us that this is a dog, but the meaning of that experience may be fear for some and love for others. Each type of experience is equally as important to acknowledge.

As depicted in Figure 3-1, *concepts* are the first level of abstraction and are defined as symbolic representations of an observable or experienced referent. Concepts help us communicate our experience and ideas to one another. Without them we would not have language.[4] In the case of the furry being, the word dog is a concept. It describes an observation shared by many. The words furry, tail, four legs, and bark are also concepts in that they are directly sensed. In experimental-type research, concepts are selected before a study is begun and are defined in such a way as to permit direct measurement or observation. In naturalistic research, concepts are derived from direct observations throughout engagement in the field.

A *construct*, the next level of abstraction, does not have an observable or directly experienced referent in shared experience. Continuing with the example of dog, for those who do not experience fear, fear is a construct. The unfearful observer of the dog may infer that others are fearful. In this case, fear is a construct because it is not directly observed or experienced by the unfearful. What may be observed are the behaviors associated with fear, that is, someone who sweats, shakes, turns pale, and moves away from the dog. These observations are synthesized into the construct of fear. Thus, fear itself is not directly observed but is surmised based on a constellation of behavioral concepts. Categories are also examples of constructs. The category of mammal or canine is an example of a construct. Each category is not directly observable, however; it is composed of a set of concepts that can be observed. Other examples of constructs include health, wellness, life roles, rehabilitation, and psychologic well-being. Each is not directly observable but is comprised of parts or components that can be observed or submitted to measurement.

So far we have been talking about single word symbols or units of abstraction. At the next level of abstraction, single units are connected and form a relationship. A relationship is defined as an association of two or more constructs and/or concepts. For example, we may suggest that the size of a dog is related to the level of fear that one experiences. This relationship has two constructs, size and fear, and one concept, dog.

A proposition is the next level of abstraction. Propositions or principles are statements that govern a set of relationships and give them a structure. For example, a proposition would suggest that fear of large dogs is caused by negative childhood experiences with large dogs. This proposition describes the structure of two sets of relationships: the relationship between size of a dog and fear, and the relationship between childhood experiences and size of dogs. It also suggests the direction of the relationship and the influence of each construct on the other.

Let us consider a theory to illustrate these different levels of abstraction and their relationship. A theory related to the furry being may look like this:

> Fear of large dogs derives from childhood experiences. It therefore can be cured by psychoanalysis that aims to reverse the negative effects of childhood experiences.

This theory explains the phenomenon "fear of dogs," is verifiable, and can lead to prediction. If, through research, we can support a positive relationship between childhood experiences and fear of dogs and the effectiveness of psychoanalysis in reducing fear, we can predict this relationship in the future and control its outcome.

Levels of Abstraction and Design Selection

We conclude our discussion of levels of abstraction with principles that relate these levels to design selection. We view abstraction as symbolic naming or the representation of shared experience. The further a symbol is removed from shared experience, the higher the level of abstraction. Each level of abstraction, as depicted in Figure 3–1, moves farther away from shared experience. The higher the level of abstraction, the more complex your design becomes. This principle is based on common sense. If the scope of your study is to examine a construct, it may not require a complex investigation such as that required for the development or testing of a full-fledged theory. Consider an example. Suppose you were interested in determining the level of cognitive recovery in a group of persons with head injury. This type of investigation is an inquiry into the construct of cognitive recovery and requires that one construct be defined and measured in your population. However, suppose you now want to move into the realm of propositions and determine the extent to which age at insult affects recovery rate. In this type of investigation, you need to define the construct "recovery" and state the nature of the relationship between it and the concept of "age." As you add more constructs, such as level of severity of injury and premorbid status, your design increases in complexity.

Let us consider another example that requires a different type of inquiry. Suppose you are interested in understanding the behaviors that place teenagers at risk for contracting the human immunodeficiency virus. You have no theory or set of constructs to investigate, so you simply collect data through asking questions and observing behaviors of the referent group. However, as your observations proceed, you find that many factors emerge that impact teenage behavior and risk. The complexity of your research approach increases as you focus on discovering specific patterns or relationships among constructs that you observe.

Although concepts, constructs, relationships, and propositions provide the basic language and building blocks of any research project, each research paradigm handles levels of abstraction differently. Let us examine the role of theory along the two continua of experimental-type and naturalistic research.

THEORY IN EXPERIMENTAL-TYPE RESEARCH

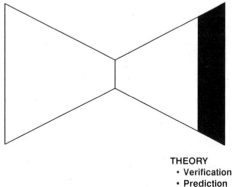

THEORY
• Verification
• Prediction

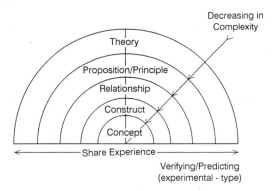

FIG 3–2.
Goal of experimental-type research: reduction of abstraction.

Experimental-type researchers begin with a theory and seek to simplify and reduce abstraction by making the abstract observable, measurable, and predictable (Fig 3–2). Usually this process involves first specifying a theory and then developing *hypotheses*. A hypothesis is linked to a theory and is a statement about the expected relationships between two or more concepts that can be tested. A hypothesis indicates what is expected to be observed and represents the researchers best hunch as to what might exist or be found based on the principles of the theory.

Let us consider an example. "Control theory" has been used extensively in medical sociology to explain deviant behavior of adolescents. Developed by Hirschi,[5] this theory posits that people with strong attachments, or "bonds," to society are less likely to participate in deviant behavior (including cigarette smoking, alcohol abuse, etc.) than those with weaker ties. As you can see, this theory identifies two basic constructs "bonds" and "deviant behavior" and suggests a relationship between these constructs. Based on this theory, a major hypothesis has been formulated and tested suggesting there is a negative association between attachment to parents and adolescent smoking regardless of the smoking behavior or attitudes of the adolescent's parents. The hypothesis suggests that the greater the attachment or bond, the less likely an adolescent will smoke. This hypothesis also suggests that such a relationship will hold regardless of the smoking behavior of the parent.

Hypotheses can be written in one of two forms: *nondirectional* or *directional*. Take for example, the following nondirectional hypothesis derived from stress theory:

For persons who care for individuals with Alzheimer's disease, there is a relationship between caregiver stress and the level of function of the care recipient.

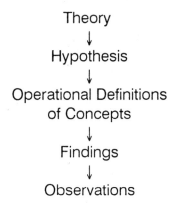

Theory

↓

Hypothesis

↓

Operational Definitions
of Concepts

↓

Findings

↓

Observations

FIG 3–3.
Use of theory in experimental-type research.

A hypothesis can also indicate the direction or nature of a relationship between two concepts or constructs, as illustrated in this example:

> Caregivers of persons with Alzheimer's disease experience increasing stress as the care recipient with Alzheimer's experiences deterioration in function.

Thus, as shown in Figure 3–3, a hypothesis is the initial reduction of theory to a more concrete and observable form. Based on the hypothesis, the researcher next develops an "operational" definition of a concept so that it can be measured by a scale or other instrument..The operational definition of the concept further reduces the abstract to a concrete observable form. An operational definition of a concept specifies the exact procedures for measuring or observing the phenomenon. In the previous example, the investigator would need to operationalize "stress" and "functional level." The process of operationalizing a concept is not easy or straightforward. Each researcher may develop his or her own way of operationalizing. This process is discussed in more detail in Chapter 14 when we examine measurement issues. Using an operational definition, the researcher makes observations that result in data that then are analyzed. Analysis determines if the findings verify, refute, or modify the theory. This scientific process, that of reducing abstraction, is used primarily to test relationships among concepts of a theory and ultimately to determine the adequacy of a theory to make predictions and thus control the phenomena under study.

The six basic characteristics of how experimental-type researchers use theory are:

- Logical-deductive process
- Primarily theory testing

- Move from theory to lesser level of abstraction
- Assumption of unitary reality that can be measured
- Knowing through existing conceptions
- Focus on measurable parts of phenomena

The process, referred to as *logical deduction,* starts with an abstraction, then focuses on discrete parts of phenomena or observations. Constructs are defined and simplified into concepts that are, in turn, operationalized into observables to allow measurement. As you can see, this process moves from a more abstract level to a less abstract or more concrete, observable level.

THEORY IN NATURALISTIC INQUIRY

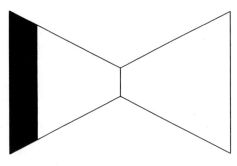

THEORY
• Theory generating

Whereas theory testing is the primary goal of the experimental-type researcher, researchers functioning within the naturalistic paradigm are most frequently, although not always, concerned with developing theory from observations. Thus, in naturalistic research, the researcher begins with shared experience and then represents that experience at increasing levels of abstraction (Fig 3–4). In this approach, definitions of concepts and constructs are not set before a study is begun but, rather, emerge from the data collection and analytic processes themselves. In many of the orientations of naturalistic research, the aim is to ground or link concepts and constructs to each observation or datum. This "theory-method link,"[6] or "grounded theory approach,"[7] is described by Glaser and Strauss[7] in this way:

> Generating a theory from data means that most hypotheses and concepts not only come from the data, but are systematically worked out

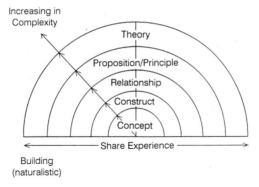

FIG 3–4.
Goal of naturalistic design: increase of abstraction.

in relation to the data during the course of the research. *Generating a theory involves a process of research* [pp 5–6].

As discussed in Chapter 2, each strategy classified as naturalistic inquiry is based on a different set of assumptions about how we come to know human experience. However, all tend to use qualitative methodologies to ground theory in observations.[6] That is, definitions of concepts and then constructs emerge from the investigation and documentation (the data) of shared experience. Based on emergent concepts and their relationships, the investigator develops theory to understand, explain, and give meaning to social and behavioral patterns. Figure 3–5 displays the basic steps through which such researchers move in the development and use of theory. Although each design structure describes this process somewhat differently, this graphic gives you the basic idea of the inductive process.

In some naturalistic inquiries, theory is used in a similar manner to experimental-type research. That is, once the investigator is emersed in observations, he or she frequently draws on well-established theories to explain these observations and make sense of what is being observed. Consider how social exchange theory, for example, may provide the ethnographer with an understanding of observed gift giving. In the process of observing gift exchange, the ethnographer may be reminded of a theory that he or she has previously read that seems to fit well with the data set. Thus, the theory is applied to the data set as an explanatory mechanism. However, the use of theory in this way occurs after the data have already been collected. Even in this way, theory does not guide data collection, nor is it imposed on the data. Rather, the data themselves suggest which theory may be relevant.

This is not to say that naturalistic inquiry is atheoretical. Theory use, however, has a different function than that in experimental-type research. All

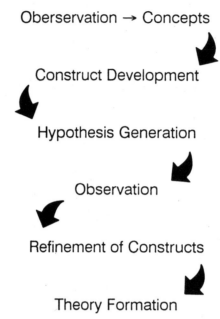

Oberservation → Concepts

Construct Development

Hypothesis Generation

Observation

Refinement of Constructs

Theory Formation

FIG 3–5.
Use of theory by naturalistic researchers.

research begins from a particular theoretical framework based on assumptions of human experience and "reality." This theoretical framework is not substantive or content specific but is structural and frames the research approach. These frameworks were discussed in Chapter 2.

In naturalistic inquiry, the researcher may also begin within a substantive theoretical framework to explore its particular meaning for a specific group of people or in specified situations (remember the Kolberg and Gilligan examples regarding moral reasoning). However, the function of the substantive theory in the naturalistic paradigm is to provide boundaries on the inquiry. The purpose is not to fit the data into the theoretical framework by reducing concepts to predefined measures. For example, in a study of an adult day-care center, Hasselkus[8] explored the application of a theoretical framework of the therapeutic use of activities to promote development, maintain health, and restore function for staff working with clients with dementia. The purpose of her study was to explore the meaning of activity in that setting. As such, although she questioned the assumptions of the therapeutic use of activities, this framework guided the type of observations she made. Her findings suggested to her that the use of a framework such as the therapeutic use of activities by staff caring for individuals with dementia may be inappropriate and lead to staff frustration.

The basic characteristics of the use of theory in naturalistic inquiry can be summarized as:

1. Primarily inductive
2. Primarily theory generating
3. Move from shared experience to higher level
4. Assumption of discovering meaning through multiple subjective understandings
5. Informant as knower
6. Focus on understanding complexity

The use of theory occurs primarily within an inductive process that moves from shared experience to higher levels of abstraction. The focus is on understanding complexity and generating theory that provides meaning to multiple subjective understandings. In this process, knowledge emerges from informants or study participants themselves.

SUMMARY

In this chapter we have made five major points:

- Theory is fundamental to the research process.
- Theory consists of four basic components that reflect different levels of abstraction. These are concepts, constructs, relationships, and propositions.
- The level at which a theory is developed and its appropriateness determine, in part, the nature and structure of research design.
- Experimental-type research moves from greater levels of abstraction to lesser levels of abstraction by reducing theory to specific hypotheses and operationalizing concepts to observe and measure as a way of testing the theory.
- Naturalistic inquiry moves from less abstraction to more abstraction by grounding or linking theory to each datum that is observed or recorded in the process of conducting research.

REFERENCES

1. Kohlberg L: *The psychology of moral development,* New York, 1984, Harper & Row.
2. Gilligan C: *In a different voice: psychological theory and women's development,* Cambridge, Mass, 1982, Harvard University Press.

3. Kerlinger FN: *Foundations of behavioral research,* ed 2, New York, 1973, Holt, Rinehart & Winston.
4. Wilson J: *Thinking with concepts,* Cambridge, Mass, 1966, Cambridge University Press.
5. Hirschi T: *Causes of delinquency,* Berkeley, Calif, 1969, University of California Press.
6. Patton MQ: *Qualitative evaluation and research methods,* ed 2, Newbury Park, Calif, 1990, Sage.
7. Glaser B, Strauss A: *The discovery of grounded theory: strategies for qualitative research,* New York, 1967, Aldine.
8. Hasselkus B: The meaning of activity: day care for persons with Alzheimer's disease, *Am J Occup Ther* 46:199–206, 1992.

Part II

THINKING PROCESSES OF RESEARCH DESIGN

Now you have a sense of the importance of research for the helping professional and how research differs from other ways of knowing about the world. We also examined the relationship between theory and research. At this point you should have a basic understanding of the different ways in which theory informs the research process along the continua of experimental-type research and naturalistic inquiry.

We define *design* broadly as the entire set of thinking and action processes that compose each step of a research effort. In the chapters that follow in Part II, we focus on the thinking processes. This effort includes problem formulation (Chapter 4), developing a knowledge base for your problem or the literature review (Chapter 5), and selection of specific methods (Chapters 7 through 11). In our discussions, we will constantly refer to how each research effort stems from a philosophical or epistemologic position, how it is based on a specific research purpose, and the role of theory in guiding the development of a project. Remember, thinking processes are planned within an epistemologic, purposeful, and theoretical context.

Each chapter in Part II introduces you to the distinct thinking processes of naturalistic inquiry and experimental-type research. You will learn about the diverse and varied ways to structure your research projects to best respond to the problems you wish to answer.

4

Framing the Problem: Question and Query

- **Identification of a Topic**
- **Framing a Research Problem**
- **Research Questions in Experimental-Type Design**
- **Research Query in Naturalistic Design**
- **Summary**
- **Exercises**

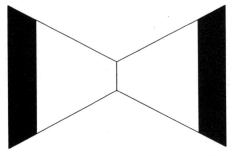

QUERY
- Broad direction indicated
- Identification of a phenomenon
- Subquestions emerge throughout the research process

QUESTION
- Concise and narrow
- Identify concepts and population

You are ready to participate in the research process, but where do you begin? The first challenge of research is the selection of a problem that is meaningful and appropriate for scientific inquiry. By meaningful we suggest that the problem should generate knowledge that is useful to the individual or to his or her profession. By appropriate we mean the selection of a problem that can be submitted to a systematic research process and that will yield knowledge to help solve the problem.

Although we routinely ask questions about and encounter problems in our professional and personal worlds, not every problem may require rigorous scientific investigation. For example, questions such as, "Why am I having difficulty developing rapport with this client? How shall I involve the family member in my intervention plan? Should helping professionals be required to wear white uniforms in a hospital setting?" can be better answered through mechanisms other than research. These questions ask for opinion and value rather than for evidence. A problem that will produce an opinion as a solution is not researchable. Or, if the question you pose can be answered by a simple yes or no, you do not need to do research to find the answer.

Most beginning researchers are able to identify a topic of interest to them. However, many find the identification of a specific research problem a very difficult task. This difficulty is caused in part by an unfamiliarity with problem formulation and its relationship to the research process. How one frames a research problem influences and shapes the development of each subsequent thinking and action process. The way a problem is framed reflects the investigator's underlying purpose in conducting the research, the investigator's preferred framework or epistemology, and the resources available to implement the project. As illustrated in Figure 4–1, the way a problem is framed will lead to the development of either a specific research question on the experimental-type continuum of research or a line of query on the naturalistic continuum or

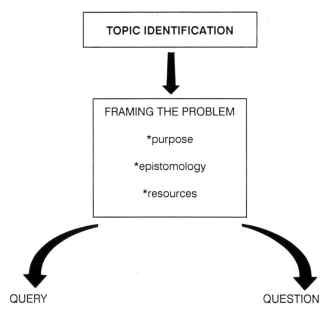

FIG 4–1.
Framing the problem: thinking sequence.

a question and query that integrates both research traditions. This chapter examines the thought process involved in identifying a topic and framing a research problem. We then discuss ways to develop meaningful and appropriate questions and queries across the full spectrum of research design.

IDENTIFICATION OF A TOPIC

The first consideration in beginning a research project is to identify a topic from which to pursue an investigation. A research topic refers to a broad issue or area that is important to a helping professional. One topic may yield many different problems and strategies for investigation. Examples of topics include post-hospital experience, health promotion, pain control, functional independence, adaptation to disability, drug abuse, hospital management practices, patient motivation, psychosocial aspects of illness, and impact of interventions on specific outcomes. It is usually easy for someone to identify a topic of interest or one that is relevant to and important for professional practice.

Where do topics and specific problem areas come from? There are five basic sources that helping professionals use to select a topic and a researchable problem:

- Professional experience
- Societal trends
- Professional trends
- Published research
- Existing theory

Professional Experience

The professional arena is perhaps the most immediate and important source from which research problems emerge. The daily ideas and confusions that emerge from cognitive dissonance or "professional challenges,"[1] like those posed at the beginning of Chapter 1, often yield significant areas of inquiry. Many of the themes or persistent issues that emerge in case review, supervision, or staff conferences may also provide investigators with researchable topics and ideas. Themes that cut across cases, such as family involvement, client motivation, and consumer perceptions of their experiences in health and human services, provide topics that may stimulate the framing of a specific research problem. For example, a rehabilitation therapist in an inpatient rehabilitation hospital posed to us an observation that she and her staff found intriguing. She observed that many of the adaptive devices for self-care that she and her staff routinely issued and spent time with in patient training were often left behind in the hospital by patients at discharge. Although she believed patients needed devices to function independently in their homes, she wondered why patients did not feel similarly.

She also wondered if the devices that were taken home were used and, if so, enhanced the individual's functioning in that environment. Discussions with therapists in other hospital sites revealed similar observations and concerns. We identified a research problem and a series of questions to examine these issues systematically in collaboration with rehabilitation therapists.[2]

Societal Trends

Social concerns and trends that are reflected in legislation and the funding priorities of federal, state, and local agencies, foundations, and corporations provide a second and critical area of potential inquiry for health and human service investigation. For example, the report *Healthy People 2000*,[3] a Public Health Service document, establishes the health related objectives and priority areas for the nation. This document provides a rich foundation for the development of research problems.

Another social arena from which research problems emerge is the set of specific requests for research generated by federal, state, and local government. Government agencies have established funding streams to provide monetary support to researchers who identify meaningful and appropriate research problems within topic areas relevant to health and human service delivery. For example, in the arena of health promotion and disease prevention, the Centers for Disease Control have issued a call for research that focuses on prevention of violent and abusive behavior and unintentional injuries. The National Institute on Alcohol Abuse and Alcoholism has requested research proposals that focus on abuse prevention for minority populations. To publicize research priorities set by the federal government, the *Federal Register,* a daily publication of the U.S. government, must list funding opportunities and policy developments of each branch of the federal government. Most libraries receive this publication, which is an invaluable resource to help identify problem areas and funding sources. The *NIH Guide for Grants and Contracts* also announces scientific research initiatives administered by the National Institutes of Health (NIH). Many announcements will specify the particular problems within a broad topic area that different branches of NIH are interested in funding.

Professional Trends

Other resources for identifying important research topics are the weekly newsletters and publications of each health and human service profession. Investigators frequently read them with the intent to determine the broad topic areas and problems of current interest to the profession. Also, professional associations establish specific short- and long-term research goals and priorities for the profession. For example, the American Occupational Therapy Foundation identified the need to develop more research in the area of therapeutic effectiveness. The Foundation has since sponsored funding competitions to support research that examines the impact of a variety of occupational therapy interventions and practices. Examining the goals and policy statements of professional associations provides a very good source from which to establish a research direction.

Research

The research world itself also provides a significant avenue from which to identify research topics and problem areas. Helping professionals encounter research through interacting with peers, attending professional meetings where research findings are reported, participating in a research project, and reviewing the literature. Reading research in professional journals provides an overview of the important studies that are being conducted in a topic area of interest. Most published research studies identify, within the text, additional research problems and unresolved issues generated by the research findings. Journals such as the *Journal of Dental Hygiene, Journal of the American Medical Association, Research in Social Work Practice, Occupational Therapy Journal of Research, Nursing Research, Medical Care, Physical Therapy, Qualitative Research in Health Care, The Social Service Review, The American Journal of Public Health,* and *Social Work Research and Abstracts* publish current research findings that are useful and relevant to the helping professions. Routinely reading such journals also provides an idea of what issues professional peers feel are important to investigate and what studies need to be replicated or repeated to confirm the findings. There are many other journals published by the helping professions and in other disciplines that may assist in the identification of a topic. Substantive topic journals that cross disciplines, such as the *Archives of Physical and Rehabilitation Medicine, Hospital and Community Psychiatry, Journal of Gerontology, Childhood and Adolescence, Topics in Geriatric Rehabilitation, The Gerontologist,* and *Journal of Rehabilitation* provide specific ideas for research studies that need to be conducted in topics relevant to helping professionals.

Existing Theory

Theories also provide an important source for generating topics and research problems. As discussed in Chapter 3, a theory posits a number of propositions and relationships between concepts. As we said, by definition,[4] to be considered a theory, each specified relationship within the theory must be submitted to systematic investigation for verification or falsification. Inquiry related to theory development is intended to substantiate the theory and advance its development or modify it by refuting some or all of its principles. Consider, for example, the well-known theory of cognitive development put forth by Piaget[5] in the mid-1920s. Piaget developed his theory of the structural development of cognition by observing his two children. Over the next 60 years, Piaget, his peers, and other scholars interested in human cognition subjected Piaget's theoretical notions to extensive scientific scrutiny. Hundreds of studies, if not more, have been conducted to corroborate, modify, or refute Piaget's principles. Further study related to Piaget's theory has been initiated to determine the application of the theory to education, health related intervention, and multiple other arenas.

FRAMING A RESEARCH PROBLEM

An endless number of problems or issues can emerge from any one topic. The challenge to the researcher is to identify one particular area of concern or specific research problem within a broad topic area. A specific research problem is a statement that identifies the phenomenon to be explored and why it needs to be examined or why it is a problem or issue. You may feel that because there are so many problems to be explored that it is very simple to select one for research. However, how one frames and states the problem is critical to the entire research endeavor and influences all subsequent thought and action processes. That is, the way in which a problem is framed determines the way in which it will be answered. Therefore, this first research step should be thought through very carefully.

To move from the selection of a broad topic area to framing a concrete research problem, the following questions can help guide your thinking:

- What about this topic is of interest to me?
- What about this topic is of relevance to my practice?
- What about this topic is unresolved in the literature?
- What about this topic is of societal or professional concern?

By answering these reflective questions, you will begin to narrow your focus and hone in on a researchable problem. Be sure, however, that whatever you develop is of interest to you. Research, regardless of the framework and type of methodology used, takes time and is thought consuming. If the problem is not challenging or exciting, the process will quickly become tedious.

Research problems range in specificity. They are either concise and narrowly defined statements along the experimental-type continuum or broadly formulated foci along the naturalistic continuum. Regardless of the level of specificity, a research problem reflects (1) the underlying purpose in conducting the research, (2) the selected framework or epistemology, and (3) resources available to implement the project. Let us discuss each of these.

Research Purpose

Based on the formulation of a problem, the researcher needs to identify the specific purpose of his or her study. The research purpose identifies why the study is being conducted. The purpose is a statement of the specific goals of a research project. In experimental-type design, the researcher is interested in describing, explaining, or predicting phenomena, as demonstrated in the following examples of purpose statements:

Descriptive. — Information on the job satisfaction of occupational therapy faculty can be helpful in the determination of whether current faculty members are likely to remain in teaching positions and what aspects of teaching are satisfying to them.[6]

Explanatory. — This study, with a relatively large national sample, examines specific coping strategies that caregivers have used to manage their day-to-day responsibilities and how the strategies are related to outcomes of caregiver well-being.[7]

Predictive. — Nursing home expenditures are the fastest rising component of personal health care expenditures and have risen recently at a rate more than double that of the consumer price index. . . . If reliable and efficient methods of predicting outcomes could be developed, it could deal with these concerns in several ways: (1) it could help in the delivery of higher quality health care through more rational allocation of scarce resources, (2) it could help in the development of better methods for assessing the quality of care, and (3) it could help in the understanding of the dynamics of nursing home outcomes.[7]

In naturalistic design, the purpose is to understand the meanings, experiences, and phenomena as they evolve in the natural setting. The following are examples of purpose statements by investigators using naturalistic design:

Ethnography. — The purpose of the study was to gain understanding of the meaning of the family caregiving experience with the hope that greater understanding would assist health professionals in working together with caregivers.[8]

Phenomenology. — The purpose of the study presented here was to evolve a structural definition of health as it is experienced in everyday life.[9]

Grounded theory. — The purpose of the present study was to identify processes used by familiar members to manage the unpredictability elicited by the need for and receipt of heart transplantation.[10]

Preferred Way of Knowing

In addition to problem formulation and one's study purpose, one's preferred way of knowing, or *epistemology,* will also influence either the development of discrete questions that objectify concepts (positivist paradigm or experimental-type design) or a query to explore multiple and interacting factors in the context in which they emerge (naturalistic paradigm) or the development of a query that integrates both approaches (which we discuss in Chapter 11). The age-old epistemologic dilemma of what is knowledge and how do we know, which we discussed in Chapter 2, is an active and forceful determinant of how a researcher frames a problem. The researcher's preferred way of knowing is either clearly articulated or implied in each problem statement. Reexamine the above purpose statements to determine the investigator's preferred way of knowing.

Resources

Resources represent the concrete limitations of the research world. The accessibility of place, group, or individuals and the extent to which time, money, and other resources are necessary and available to the researcher to implement the question or query are examples of some of the natural constraints of conducting research. These real-life constraints actively shape the development of the research problem an investigator pursues and the scope of the project that can be implemented.

Once a research problem and study purpose have been framed, a specific question or query can be developed. Let us examine how questions are developed for experimental-type design and then see how a problem evolves into a query in naturalistic research.

RESEARCH QUESTIONS IN EXPERIMENTAL-TYPE DESIGN

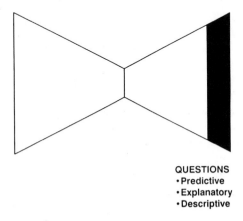

QUESTIONS
• Predictive
• Explanatory
• Descriptive

In experimental-type design, the researcher identifies a broad topic, specifies a problem area within the topic, and then articulates a clear question that guides the investigation. The question is concise and narrow and establishes the boundaries or limits as to what concepts, individuals, or phenomena will be examined. The question is posed a priori to engaging in the research process and forms the basis from which all other research action processes emerge. The purpose of the research question on the experimental-type continuum is to articulate not only the concepts but the structure and scope of the concepts that will be studied. Research questions lead the investigator to specific design and data collection action processes that involve some form of observation or measurement to answer the question. Questions are developed deductively from the theoretical principles that exist and are presented in the literature. Questions seek to either (1) describe a phenomenon, (2) explore relationships among phenomena, or (3) test existing theory or models. These question types reflect three different levels of knowledge and theory development about the topic of interest, ranging from level 1, where little to nothing is known about a topic, to level 2 where descriptive knowledge is known but relationships are not yet understood, to level 3 in which there is a substantial body of knowledge and development of well-defined theory. Let us examine each question type in more detail.

Questions That Seek to Describe A Phenomenon

Questions that aim to describe phenomena are referred to as Level 1 questions.[11] These questions are designed to elicit descriptions of a single topic or a single population about which little or nothing is known. Level 1 questions lead to exploratory action processes where the intent is to describe. A Level 1 question is often necessary if a body of knowledge has been developed for one population but not in another. For example, although a large body of knowledge exists regarding exercise capacity in men, little is known about the phenomenon in women. Therefore, research based on a Level 1 question would be meaningful and appropriate to describe exercise capacity and normal values and ranges for women.

Level 1 questions focus on describing the nature, extent, or direction of one concept or *variable* in a population. To describe a variable, which can be defined as a characteristic or phenomenon that has more than one value, one must operationalize or define it in such a way as to permit its measurement. Thus, Level 1 questions focus on measuring the nature of a particular phenomenon in the population of interest.

The example given earlier regarding the use of adaptive devices illustrates this level of questioning. Before the study was implemented, a literature review was conducted by the investigators and yielded a limited body of knowledge regarding use patterns of adaptive equipment over time, specifically for older rehabilitation clients. There was also no theoretical framework to explain adaptive equipment use patterns. Based on the limited literature and nonexistent theory development in their area, Gitlin et al.[2] designed a Level 1 descriptive study to describe adaptive equipment use in the home of rehabilitation clients and to develop an understanding of reasons for use and nonuse by older clients with a range of disabilities. The specific research questions were:

- What is the frequency of use of adaptive devices by older adults discharged to their home from rehabilitation?
- What are the reasons for use and nonuse of devices in the home?

In these research questions, the main concept or variable measured was adaptive equipment use. The population of interest was older adults with mixed disabilities. The focus of the level 1 questions was on "the what," as can be seen in the wording of the questions. Some examples of other Level 1 questions include:

- What are the premorbid psychologic characteristics of individuals who experience traumatic brain injury?
- What are the value orientations of occupational therapists?
- What is the level of interest that social work students express in working with persons with serious mental illness?

Each of these questions uses the stem of "what are" or "what is" and refers to one population. In these examples, one population is identified such as individuals with traumatic brain injury, occupational therapists, and social work students. Also, one concept is examined in each study: premorbid psychologic characteristics, value orientation, and interest. Each of these concepts can be measured or empirically examined. In essence, Level 1 questions describe the parts of the whole. Remember that the underlying thought process for experimental-type research is to come to know about something by examining its parts and their relationships. Level 1 questioning is the foundation step for clarifying the parts and their nature. In the scheme of levels of abstraction (remember our discussion in Chapter 3), Level 1 questions address the lowest levels of abstraction, those of concepts and constructs.

As we will discuss in Chapters 7 and 8, these Level 1 questions lead to the development of descriptive-type designs such as surveys, exploratory or descriptive studies, trend designs, feasibility studies, need assessments, and case studies.

Questions That Explore Relationships Between Phenomena

This research question, Level 2, builds on and refines the results of Level 1 studies. Once a "part" of a phenomenon has been described and there is existing knowledge about it in the context of a particular population, the experimental-type researcher may now begin to pose questions that are relational. The key purpose of this level of questioning is to explore relationships among phenomena that have already been studied at the descriptive level. Here the stem question asks, "What is the relationship?" and the topic contains two or more concepts or variables. For example, a researcher might be interested in asking, "What is the relationship between exercise capacity and cardio-vascular health in middle-aged men?" In this case the two identified variables measured are "exercise capacity" and "cardiovascular health." The specific population is middle-aged men. As you can surmise, Level 1 research must have been accomplished for the two variables to be defined and operation-alized.

Let us return to the example of the last three questions. Suppose we have conducted studies to address these three level 1 questions. If we were to continue our research agenda in each of these areas of inquiry, we would be ready for the following Level 2 questions:

- What is the relationship between learning disabilities and incidence of traumatic brain injury?
- What is the relationship between the value orientations of occupational therapists and practice techniques used in acute care settings?
- What is the relationship between previous work experience with persons with serious mental illness and social work student interest in working with this population?

Referring to the levels of abstraction that we discussed previously, Level 2 questions address the level of relationships. This represents the next level of complexity from Level 1 questions. These questions continue to build on knowledge in the experimental-type framework by examining not only parts but also their relationships and may also explore the nature and direction of these relationships. Level 2 questions lead primarily to research using passive observation design, discussed in detail in Chapter 8.

Questions That Test Knowledge

Level 3 questions build on the knowledge generated from research conducted at Levels 1 and 2. These questions ask about a cause and effect relationship between two variables to test knowledge or the theory behind the knowledge. At Level 2 we asked the question, "What is the relationship between the value orientations of occupational therapists and practice techniques used in acute care settings?" Suppose in that study it was revealed that occupational therapists who valued the inclusion of client and caregiver goals and values in the therapeutic process were able to build better rapport and had more positive outcomes than those who did not own that value orientation. Building on this knowledge, you would be able to ask the Level 3 question, "Why is inclusion of client and caregiver perspectives effective in promoting client outcomes?" The resultant study would test the theory behind why such a therapeutic approach leads to improved client outcomes. At this level of questioning, the purpose is to predict what will happen and provide a theory to explain why. Based on a Level 3 question, specific predictive hypotheses (statements predicting the outcome on one variable based on knowing about the other) are formulated.

In a Level 3 question, it is assumed that two concepts are related based on previous research findings (from Level 2 research). The point of a study at Level 3 is to test these concepts in action by manipulating one to effect the other. Level 3 questioning is the most complex of experimental-type questioning. Once the foundation questions formulated at Levels 1 and 2 are answered, Level 3 questions can be posed and answered to develop knowledge not only of parts and their relationships but of how these parts interact and why they interact to cause a particular outcome. Level 3 questions examine higher levels of abstraction, including principles, theories, and models.

To answer a Level 3 question, the research action processes must be capable of revealing causal relationships among variables. A "true experimental" design or variation of it can be implemented to support the effectiveness of including client and caregiver perspectives in treatment. (See Chapters 7 and 8 for a full explanation of hypothesis and true experimentation.)

Table 4–1 summarizes our discussion thus far.

TABLE 4–1.
Summary of Question Types in Experimental-Type Research

Level	Stem	Level of Abstraction	Design Possibilities
1	What is What are	Concepts and Constructs	Survey Exploratory Descriptive Case study Needs assessment
2	What is the Relationship	Relationships	Survey Correlational/passive Observation Ex post facto
3	Why	Principles Theories Models	Experimental designs Quasi-experimental

Brink and Wood[11] provide some helpful rules to follow in developing an appropriate level of question:

1. At Level 1, have only one variable and one population in the topic.
2. At Level 2, a minimum of two variables is needed.
3. If there is a cause or effect to be investigated, write the question at Level 2 or 3.
4. If the words "cause," "effect," or any of their synonyms appear in the question, either eliminate those words or specify what they are and how they vary.
5. All variables must be written so that they vary.
6. At Level 3, there must be two variables that specify a cause and an effect.
7. If a Level 3 question is written, make sure it is both ethical and possible to manipulate the causal variable. If not, rewrite the question at Level 2.

RESEARCH QUERY IN NATURALISTIC DESIGN

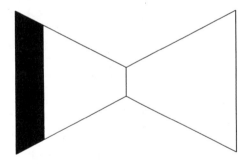

QUERY
•Formulation and reformulation
•Subquestions emerge in the field

As discussed previously, there are a range of philosophical perspectives that inform inquiry within naturalistic research. Each perspective in this paradigm is rooted in a distinct philosophical tradition and approaches the task of framing the problem somewhat differently. However, there are some broad similarities in approaches to problem development that we will discuss here.

Researchers on the naturalistic continuum generally identify a topic and a broad problem area or phenomenon to be studied that then becomes the field from which a query is pursued. We use the term "query" to refer to a broad statement that identifies the phenomenon or natural field of interest. "Phenomena" may refer to the symbolic patterns of interaction in a cultural grouping, the experience of disability, the meaning of pain, and so forth. The natural field in which the phenomena occur forms the basis for discovery in which more specific and limited questions evolve in the course of conducting the research. That is, as new insights and meanings are obtained, the initial problem statement and query become reformulated. Based on new insights and issues that emerge in the field or the action of research, the investigator formulates smaller, concise questions, which are then pursued. These smaller questions are contextual in that they evolve inductively within the study itself from the investigator's ongoing efforts to understand the broad problem area. This interactive questioning–data gathering–analysis–reformulating of questions represents the core process of naturalistic inquiry. Let us look at three qualitative methods within the naturalistic paradigm to see how each develops a researchable query and engages in a process of formulation and reformulation.

Ethnography

As the primary research approach in anthropology, ethnography is concerned with the description and interpretation of cultural patterns of groups and the understanding of the cultural meanings people use to organize and interpret their experiences.[12] In this approach, the researcher assumes a "learning role" in which different cultural settings are experienced firsthand to interpret, make sense out of, and bridge the world and culture of the researcher to that of the researched.[13] Once the ethnographer has identified a phenomenon and cultural setting, a line of query is pursued. Agar[13] describes the broad nature of the initial questioning of an ethnography as such:

> When you stand on the edge of a village and watch the noise and motion, you wonder, "Why are these people and what are they doing?" When you read a news story about the discontent of young lawyers with their profession, you wonder, "What is going on here [p. 13]?

As you can see, experimental-type questions and hypotheses stand in stark contrast to the interpretative opened query posed by the ethnographer.

As the processes of data gathering and analysis proceed in tandem, specific questions emerge and are pursued. These questions emerge in the field as a consequence of what Agar[13] labels as "breakdowns," or disjunctions. He[13] describes this occurrence in this way: "A breakdown is a lack of fit between one's encounter with a tradition and the schema-guided expectation by which one organizes experience" [p. 21]. Put simply, breakdowns represent differences between what the investigator expects and what the investigator observes. These differences stimulate a series of questioning and further investigation. Each subquestion is related to the broader line of query and is investigated to resolve the breakdown and develop a more comprehensive understanding of the phenomenon in its entirety.

An ethnographic query, therefore, establishes the phenomenon, setting of interest, or both. The query also sets up the thought and action processes necessary to understand the boundaries of a culture and the human experience that occurs within those boundaries. To summarize, once a query has been posed, questioning occurs simultaneously with collecting information and making sense of it. One process drives the other. The interactive questioning, data gathering, and analytic processes result in the reformulation and refinement of the problem and the structuring of smaller questions. These questions are then pursued in the field to uncover underlying meanings and cultural patterns.

Helping professionals have used ethnographic methods such as interviewing and participant observation, discussed more fully in Chapters 9 and 10, to examine cultural variations in response to disability, adaptation, health services utilization, health care practices, and so forth. The health or human service professional conducting an ethnography may start with a general query, such as "How do children with terminal cancer express their fear and pain?" or "How are individuals with severe disabilities able to live independently and what are their patterns of adaptation?"

Phenomenology

The purpose of this line of inquiry is to uncover the meaning of how humans experience phenomena through the description of those experiences as they are lived by individuals.[14] The first research step from this perspective, then, is to identify the phenomenon of investigation. Being in pain, homeless, or sad may be phenomena that are of relevance to the helping professions. From the articulation of a phenomenon, a research query is generated, such as "What is the meaning of being homeless for middle-aged women?" "What is the meaning of fear for persons with traumatic injury?" "What are the common elements in experiencing a feeling of well-being among the elderly? Different from ethnography, phenomenologic queries focus on experience from the perspective of individuals and do not seek to broaden that experience to understand a group or cultural context. However, similar to ethnography, the research begins with a broad query. Based on what the investigator comes to know through the action of research, subquestions and other inquiries are developed that further inform the meaning of experience for the individuals participating in the study.

Grounded Theory

Grounded theory is a method in naturalistic research used primarily to generate theory using an inductive process.[15] The researcher begins with a broad query in a particular topic area and then collects relevant information about that topic. As the action processes of data collection continue, each piece of information is reviewed, compared, and contrasted with other information. From this "constant comparison" process, commonalities and dissimilarities among categories of information become clear, and principles, and ultimately theory to explain observations, are inductively developed. Thus, the queries for grounded theory relate not to specific domains but to the structure of how the researcher using grounded theory wants to organize his or her findings. For example, questions that would be answered through a grounded theory approach are:

- What theoretical principles characterize the experience of women who become homeless?
- What similarities and differences can be revealed among ways in which traumatically injured patients experience their acute care hospitalization?
- What theoretical principles explain why first-year college students select academic majors that will prepare them for a career in a helping profession?

As you can see in these questions, each indicates that the research aim is to reveal theoretical principles about the phenomenon under study.

Grounded theory can also be used to modify existing theory or to expand on what is known in different ways. Queries related to these research objectives include:

- How can current theory of moral development be expanded or modified to explain the moral development of Native American children and youth?
- What is the relevance of current theory on career development in male children to female children?

In these two questions, grounded theory would be structured to address current theory from a new and inductive perspective.

As you can see, ethnography, phenomenology, and grounded theory reflect distinct approaches in naturalistic inquiry. Queries developed from each design approach reflect a different purpose, preferred way of knowing and are shaped by the particular resources available to the investigator.

SUMMARY

There are many ways to frame a research problem. Each approach to problem formulation and query and question development differs along the two continua of research design. In experimental-type designs, the question drives each subsequent research step. In naturalistic inquiry, the query sets the initial boundaries for the study by becoming reformulated in the process of data collection and analysis.

Let us consider an example of an investigation into the phenomenon of caregiving to highlight these differences in approaches. There are many issues one can investigate under the broad rubric of caregiving. For example, if one's purpose is to understand the relationship between the act of caregiving and caregiver sense of well-being, one may pose a specific question such as, "What is the relationship between care management style and sense of self-efficacy in caregivers of spouses with the diagnosis of dementia?" What type of question is this? If you guessed level 2 along the experimental-type continuum, you are correct. In this question a specific population is identified, that of spouses caring for individuals with a diagnosis of dementia. This question narrows the area of concern of relationships between caregiver style and caregiver outcomes to two constructs: self-efficacy or the perception of one's own abilities, and care management style, or one's approach to managing multiple care responsibilities. Here the investigator has isolated constructs based on a deductive thinking process and an orientation to knowing which favors experimental-type design. The constructs posed in the research question may have been identified from reading the research literature, from practice experience, or from federal funding initiatives that isolate self-efficacy and management as important constructs from which to understand caregiver experiences. The question would lend itself to a design that involves interviewing a substantial number of caregivers to examine the stated relationships. The way the question is stated assumes that the investigator has the necessary fiscal resources and personnel to implement the research action processes to answer the question.

In contrast, if one's purpose is to understand the experience of caregiving and one's preferred epistemologic framework is based in a naturalistic paradigm, a research query initially posed as a broad statement might be developed. For example, the query "How do caregivers experience the act of caring for their spouses with dementia in the home?" identifies caregiving spouses in the home as the field of query and the personal and lived experience of caring as the particular phenomenon of interest. In such a study, relevant constructs will emerge based on uncovering and revealing the meaning of caregiving from caregivers themselves. The initial formulation of the problem as "caregiver experience" assumes that the investigator does not know the range of such experiences. The investigator proceeds inductively based on the framework that there are multiple caregiver experiences and realities that need to be discovered and understood. It is from understanding these realities that meaningful concepts, constructs, and relationships among constructs will emerge in the course of the study. In this case, the monetary and personnel resources available to the investigator will determine the scope of the field that is investigated and the time engaged in the field observing and interviewing caregivers.

In summary, in experimental-type design, refinement of a research question occurs before any action processes. Conciseness and refinement are critical to the conduct of the study and are the hallmarks of what makes the research question meaningful and appropriate. Experimental-type questions are definitive, structured, and derived deductively before one engages in research actions. In contrast, for naturalistic design, such question refinement emerges from the action of conducting the research. The development of questions occurs inductively and emerges from the interaction of the investigator with the field or phenomenon of the study. Thus "questions" have quite different levels of meaning and implication for the conduct of the study in experimental-type design and naturalistic research. Research queries and subquestions are dynamic, ever changing, and derived inductively.

EXERCISES

1. To test your understanding of the differences between levels of questioning in experimental-type designs, select a problem area in which you are interested. Then frame the problem in terms of a level 1, 2, and 3 question.

2. Now take the same problem area and formulate a line of query to pursue with a naturalistic design.

3. Review your different problem statements and specific query or questions. Identify the different assumptions each makes about level of knowledge, preferred way of knowing, and required resources to conduct the study. Use the grid that follows to assist you.

4. Select three research articles in the literature and identify each research question or query. After you have identified them, characterize the nature of each using the grid.

Question/Query	Assumption of Level of Knowledge	Preferred Way of Knowing	Required Resources
1.			
2.			
3.			

REFERENCES

1. Schon D: *The reflective practitioner,* New York, 1983, Basic Books.
2. Gitlin LN, Levine R, Geiger C: Adaptive device use in the home by older adults with mixed disabilities, *Arch Phys Med Rehabil* Feb 1993; 149–152.

3. *Healthy people 2000*, Washington DC, 1990, US Government Printing Co.
4. Wilson J: *Thinking with concepts*, Cambridge, Mass, 1966, Cambridge University Press.
5. Piaget J [Warden M, translator]: *Judgment and reasoning in the child*, New York, 1926, Harcourt, Brace.
6. Rozier C, Gilkeson G, Hamilton BL: Job satisfaction of occupational therapy faculty, *Am J Occup Ther* 45:160–165, 1991.
7. Palmore EB: Predictors of outcome in nursing homes, *J Appl Gerontol* 9:1172–1184, 1990.
8. Hasselkus BR: The meaning of activity: day care for persons with Alzheimer disease, *Am J Occup Ther* 46:199–206, 1992.
9. Parse RR, Coyne AB, Smith MJ: Nursing research traditions: qualitative and quantitative approaches. In *Nursing research, qualitative methods*, Bowie, Md, 1985, Brady Communications, pp 1–8.
10. Mishel M, Murdaugh C: Family adjustment to heart transplantation: redesigning the dream, *Nurs Res* 36:332–338, 1987.
11. Brink PJ, Wood MJ: Basic steps in planning nursing research: from question to proposal, ed 3, Boston, 1988, Jones & Bartlett.
12. Spradely J: *Participant observation*, New York, 1980, Holt, Rinehart & Winston.
13. Agar M: *Speaking of ethnography*, Newbury Park, Calif, 1986, Sage.
14. Farber M: *The aims of phenomenology: the motives, methods, and impact of Husserl's thought*, New York, 1966, Harper & Row.
15. Glaser B, Strauss A: *Grounded theory: strategies for qualitative research*, New York, 1967, Aldine.

5

Developing a Knowledge Base Through Review of the Literature

- **Why Review the Literature?**
- **How to Conduct a Literature Search**
- **Summary**
- **Exercises**

Once you have an idea of a research problem, going to the literature will help you refine your thinking and move it along to frame a specific query or question. All of us probably have recollections of long hours at the library poring through the literature to discover what others have written about the topic in which we are interested. Reviewing the literature is one of the most exciting and significant aspects of doing research. However, it is often a misconceived and undervalued activity.

A literature review in the world of research serves a slightly different purpose than a review you may have conducted for other types of activities or classes. It is a process in which the researcher *critically* reviews literature that is directly or indirectly related to both the topic and the proposed strategy of conducting the research. The information you obtain from a critical review directs you to a particular research focus and strategy. Finding out what others know and how they come to know it helps determine how your research fits within the body of existing knowledge and the unique contributions of your research to the scientific enterprise.

In this chapter, we begin with a discussion of the different reasons for conducting a literature review. We then present specific steps in conducting a literature review. There is often so much literature directly related to a topic and so many other related bodies of research that the review process may feel very overwhelming. However, there are tricks of the trade, so to speak, to facilitate a systematic and comprehensive review process that is manageable and even enjoyable.

WHY REVIEW THE LITERATURE?

You need to review the literature for four major reasons. These are to *critically* determine:

- What previous research on the topic has been done.
- The level of theory and knowledge development relevant to your problem area.
- The relevance of the current knowledge base to your problem area.
- A rationale for the selection of your research strategy

Let us examine each of these reasons in more depth.

Determining What Research Has Been Done on the Topic of Inquiry

Why conduct a study if it has already been done and done to your satisfaction? The key here is "to your satisfaction." To determine if the current literature is sufficient to help you solve a professional problem, you must critically evaluate how others have struggled with and resolved the same question. An initial review of the literature provides you with a sense of what has been done in your area of interest. The review helps you identify current trends and ways of thinking about your topic, contemporary debates in your field, gaps in the knowledge base, the way in which the current knowledge on your topic has been developed, and the conceptual frameworks used to frame your problem.

Sometimes an initial review of the literature will steer you in a different research direction from that which you had originally planned. For example, let us say you want to examine the extent to which mild aerobic exercise promotes cardiovascular fitness in the elderly. You find several studies that already document a positive outcome. However, many of these authors suggest that further inquiry is necessary to determine the particular exercises that promote cardiovascular fitness, as well as those that maintain joint mobility. Although this area may not be what you originally had in mind, the literature review refines your thinking and directs you to specific problem areas that need greater research attention. You will need to decide if replication of the reported studies is important to verify study findings or whether your focus should be modified according to the recommendations in the literature.

Let us consider another example of how literature influences one's research direction. Kaplan[1] was interested in examining the knowledge of clinical social workers and psychologists in regard to their clients' ethnic backgrounds. Initially she hypothesized that clinicians who were knowledgeable about a client's ethnicity would deliver more effective counseling services. When she went to the literature, she found only position papers from professional societies as to the importance of a multicultural perspective in practice. There was no research to support the relationship she had hypothesized. Also, she was unable to locate any study or instrument that had measured "ethnic knowledge." She therefore decided to modify her original research question and conduct a study to develop and test an instrument to measure this construct, as well as to determine therapists' knowledge of specific cultures.

Determining Level of Theory and Knowledge Development Relevant to Your Project

As you review the literature, not only do you need to describe what exists, but most important, you need to *critically* analyze three important related factors in each work:

* Level of knowledge
* How the knowledge has been generated
* Boundaries of the study

Level of Knowledge

First, when you read a study, evaluate the level of knowledge that emerges from the study. Remember the discussion in Chapter 3 on theory? As we indicated in that chapter, studies produce varying levels of knowledge from descriptive to theoretical. The level of theory development and knowledge in your topic area is a major determinant of the type of research strategy you will select. As you read related studies in the literature, think about the level of theoretical development that exists. Is the body of knowledge descriptive, explanatory, or at the level of prediction?

How Knowledge is Generated

Once you have determined the level of knowledge that exists, you need to evaluate how that knowledge has been generated. In other words, you need to identify the research strategy or design that is used in each study. As you identify the design, critically examine whether it is appropriate for the level of knowledge the authors indicate exists in the literature and whether the conclusions of the study are consistent with the design strategy.

We all know that there is a tendency to read the introduction to a study and then jump to the conclusions to see what it can do for you. However, it really is important to read each aspect of a research report, especially the section on methodology. As a matter of fact, you will be surprised to find that, unfortunately, the literature is full of research that demonstrates an inappropriate match between research question, strategy, and conclusions. As you read a study, ask yourself three interrelated questions:

- Are the research methods congruent with the level of knowledge reviewed by the investigators in their report?
- Do the research procedures adequately address the proposed research question?
- Is there compatibility or a fit between procedures, findings, and conclusions?

The answers to these questions will lead you to understand and critically determine the appropriateness of the level of knowledge that has been generated in your topic area.

Boundaries of Study

It is also important to determine the boundaries of the studies that you read and their relevance to your study problem. By boundaries we mean the "who, what, when, and where" of a study. For example, suppose you are interested in testing the effectiveness of a hypertension reduction program in a rural setting. Let us say previously published studies have reported positive outcomes for a range of interventions. However, on closer examination, you see that these studies have been conducted in urban areas and included participants who were primarily white and male and have backgrounds very different from the population you plan to include in your study. Therefore, it would be important to examine studies that tested other types of health promotion interventions in rural settings and included participants with characteristics similar to those your study intends to target. By extending the literature review in this way, you would obtain a better understanding of the issues and design considerations that are important for the specific boundaries of your study. Also, you may discover that the level of theory and knowledge development in hypertensive risk reduction programs is quite advanced for one particular population, let us say, white urban males, but very undeveloped in respect to another population, such as rural black females.

If you find no literature in your topic area, you might identify a body of literature that is somehow analogous to your study as a way of directing how you develop your strategy. For example, in a study of clinical mastery in occupational therapy, DePoy[2] found relatively little to no research conducted in this area. Yet, studies of excellence and achievement had been conducted in fields such as business and music. DePoy reviewed these bodies of literature to shape her methods.

Determining the boundaries of the literature helps you to evaluate the level of knowledge that exists for the particular population and setting you are interested in studying.

Determining Relevance of the Current Knowledge Base to Your Problem Area

Once you have evaluated the level of knowledge advanced in the literature, you need to determine its relevance to your idea.

Let us say you are thinking of conducting an experimental-type study. In your literature review, you must find research that points to a theoretical framework relevant to your topic and research that identifies specific variables and measures for inclusion in your study. Your literature review must also yield a body of sufficiently developed knowledge so that hypotheses can be derived for your study.

For example, in a study of attrition in occupational therapy, Bailey[3] reviewed literature in four areas of work: job dissatisfaction, economic factors, family responsibility, and other factors. Based on an extensive review, Bailey found that the level of theory development on attrition was sufficient to support an experimental-type strategy. Also, she justified the selection of specific variables and instruments based on the literature.

If your literature review shows a clear gap in existing knowledge or a poor fit between a phenomenon and current theory, you probably should consider a strategy on the naturalistic continuum. Willoughby and Keating's[4] study on caregiving illustrates how investigators determined the relevance of existing literature to their particular study problem. These investigators were interested in examining how people care for a relative with Alzheimer's disease. Although there is an abundance of caregiver literature, they[4] were unable to find studies that "discuss how a caregiver moves from beginning to subsequent stages of phases of caregiving or whether there are distinct events or understandings that separate this from subsequent stages" [p. 29]. They[4] also noted that from a methodologic standpoint, "Few studies of caregiving have been grounded in the experiences of caregivers themselves" [p. 30]. Based on their review, the authors questioned the relevance of the literature to the problem area. They chose a naturalistic strategy to ground an understanding of caregiving in the experiences and perspectives of caregivers. The use of a naturalistic strategy was supported by the absence of a well-developed theory of caregiving and the absence of research that examined the process over time.

Providing a Rationale for Selection of the Research Strategy

Once you have determined the content and structure of existing theory and knowledge related to your problem area, the next task is to determine a rationale for the selection of your research. A literature review for a research grant proposal or research report is written to directly support your research query or question and choice of design. You must synthesize your critical review of existing studies in such a way that the reader sees how your study is a logical extension of current knowledge in the literature.

First, let us see how to conduct a literature search; then we will offer guidelines for developing a written rationale.

HOW TO CONDUCT A LITERATURE SEARCH

Conducting a literature search and writing a literature review are exciting and creative processes. However, there may be so much literature, that if you do not know how to go about these tasks, it can initially seem overwhelming. In this section, we identify six steps to this process to guide you in searching the literature, organizing the sources, and taking notes on the references that you plan to use in a written rationale:

 Step 1: Determine when to do a search
 Step 2: Delimit what is searched
 Step 3: Access data bases for periodicals, books, and documents
 Step 4: Organize the information
 Step 5: Critically evaluate the literature
 Step 6: Write the literature review

Step 1: When to Do a Search

The first step in a literature review is to determine when such a review should be done. In studies along the experimental-type continuum, a literature review always comes first and precedes the final formulation of a research question and the implementation of the study.

In studies along the continuum of naturalistic inquiry, the literature may be reviewed at different points in time throughout the project. Because the literature is not critical for defining variables and instrumentation, it serves other purposes. When one reviews the literature depends on the purpose it will serve. Sometimes, for example, the literature is used as an additional source of data and is included as part of the information gathering process that is subsequently analyzed and interpreted. Another purpose of the literature review in naturalistic inquiry is to inform the direction of data collection once the investigator is in the field. As discoveries occur in data collection, the investigator may turn to the literature to obtain some guidance about how to interpret emergent themes and how to proceed in the field.

To the extreme are forms of research in which no literature is reviewed before or during fieldwork. In a study of prison violence using an "endogenous" research design[5] (discussed in Chapter 10), the researcher and the research team of prison inmates did not believe that a review was necessary or relevant to the purpose of their study.

However, for the most part, researchers working along the continuum of naturalistic inquiry review the literature before conducting research to confirm the need for a naturalistic approach. For example, in the caregiver study described earlier, Willoughby and Keating[4] reviewed the literature before framing the research strategy, and their review supported a naturalistic strategy.

Keep in mind that although an extensive review of the literature refines the research question/query and design, investigators are always updating their literature review throughout the conduct of a study.

Step 2: Delimiting the Search

Once you have made a decision about when to do a literature review, your next step involves setting parameters as to what is relevant to search. After all, it is not feasible or reasonable to review every single topic that is somewhat related to your problem. Delimiting or setting boundaries to the search is an important yet difficult step. The boundaries you set must assure a comprehensive review but one that is practical and not overwhelming.

One useful strategy used by investigators along both design continua is to base a search on the concepts and constructs that are contained in the initial question or query. For example, suppose you want to determine whether participation in an exercise program for elderly nursing home residents improves perceived health status. The concepts and constructs contained in this initial formulation include exercise, nursing home residents, and perceived health status. You would begin a search using these variables to identify literature and determine how these have been studied.

Step 3: Mechanics of a Search

Now that you determined when the literature search will take place and have established some limits on what will be searched, you are ready to go to the library. Over the past 10 years, library searches have become technologically sophisticated and are a great advantage over conventional searching mechanisms. However, the most valuable resource for finding references in the library is still the *reference librarian*. Working with a good librarian is the best way to learn how to navigate the library and the numerous data bases that open the world of literature to you.

Once in the library, you have a choice about how to begin a search. You can access literature through *on-line (computerized) data bases*, through the *card catalogue*, or through other *indexes and abstracts*. In health and human service research you will most likely want to search three categories of materials: books, journal articles, and government documents. You may also find newspapers and newsletters useful, especially those from professional associations. In addition, you can construct a literature search based on the references of research studies you have obtained. However, do not just depend on what these authors identify as important. First, references from other articles may not represent the broad range of studies you probably should review, nor will they represent the most current literature. There is sometimes a delay of 1 year or more between the submission of a manuscript and its publication date.

Searching Periodicals and Journals

Usually most researchers begin their search by examining periodicals and journals. To begin your search, it is most helpful and time efficient to work on the on-line data bases. Searching data bases can be fun, but you must know how to use the systems. You can search the data bases by examining subjects, authors, or titles of articles.

Let us first consider a search based on subject. Start out by identifying the major constructs and concepts of your initial query, then translate these into what is referred to as "key words." Most journals included in on-line data bases require authors to specify and list key words that reflect the content of their study. The data bases group these studies according to these key words. For example, Figure 5–1 presents part of the title page of Bailey's article on attrition in occupational therapy. In addition to the title, author, and abstract, there are "key words."

You can obtain help from the librarian in formulating key words, or the computer will help direct you to synonymous terms if the words you initially select are not used as classifiers. Most on-line systems are user friendly and interactive in that the computer itself provides you with step-by-step directions on what to do.

If you were to search any of the key words listed in Bailey's article, you would come across a citation of the article.

Let us return to the example of the exercise study to show the importance of key words. We will begin our search with the key words "exercise" and "elderly." The combination of terms is critical in delimiting and focusing your search. For example, when we searched "Carl's Uncover," a data base of journal and newspaper articles available in many university libraries, we first entered the key word "exercise." The computer then counted the citations that were relevant to the subject of exercise and indicated that there were 4,495 different sources in the past 5 years. Such a review of all of these sources would be too cumbersome to say the least. We then entered our second key word, "elderly," at a prompt that requested further delimiting. The computer then searched for works containing both topics and indicated a total of 46 articles.

After the computer indicates how many citations are found, usually the next step is to display the sources. Often you have a choice to examine only the citation or the abstract as well. If your on-line system is connected to a printer, you can select a print command and print out the citation and abstract.

Reasons for Attrition From Occupational Therapy

Diana M. Bailey

Key Words: allied health personnel • occupational therapists, manpower

This study examined the reasons why occupational therapists have left the field of occupational therapy. The purpose of the study was to find ways to prevent attrition and to bring back those who have left the field as a means to address the profession's personnel shortage. Questionnaires from 696 therapists who have left the profession were analyzed. The therapists' most common reasons for leaving were (a) childbearing and child rearing; (b) geographic relocation and subsequent inability to find a job; (c) excessive paperwork; (d) desire for increased salary and promotional opportunities; (e) high caseloads, stress, and burnout; (f) the actual practice of occupational therapy not being what was expected; (g) dissatisfaction with bureaucracy; (h) the chronicity and severity of the clients' illnesses; and (i) an inability to find part-time work. Most therapists who left the profession did not return to practice because they felt professionally out of date and unable to compete with younger therapists.

FIG 5–1.
Title page, including "Key Words." (From Bailey DM: *Am J Occup Ther* 4:23–29, 1990.)

You may also search the system by entering authors' names or words that you think may appear in the titles of sources. This entry expands your search to relevant articles that may not have included the combination of key words you initially used. The computer will search the references and compile a list of sources you then can choose to obtain.

In some cases it is more efficient to ask the librarian to conduct a search for you. Usually large data bases such as ERIC (Educational Information Resources Information Center) and Medline are accessed in this manner. Asking the librarian to conduct an evening search will be less expensive, because the telephone lines that are used to connect the library's computer to the central data base charge less in off-peak hours.

It is also possible to conduct a more limited search by accessing data bases that index or abstract studies in only a specific field or subject. For example, many libraries now have a data base called "PsychLIT," which contains sources from the field of psychology and related areas. "Mini-Medline" is another limited data base frequently available in universities with health and medical departments. Each of the data bases gives user friendly directions for access and use.

You also may want to do a manual search in the important indexes and abstracts to ensure that you have fully covered the literature relevant to your study. In health research, useful indexes and abstracts include:

- *Social Science Citation Index*
- *General Science Index*
- *Reader's Guide to Periodical Literature*
- *Index to Government Periodicals*
- *New York Times Index*
- *Psychological Abstracts*
- *Medical Psychological Abstracts*
- *Cumulative Index to Nursing and Allied Health*

You can search these volumes by author, subject, or title, just as you have done with the on-line data base. Of course, there are so many abstracting and indexing services that it would be impossible to list all of the relevant ones here. However, *Ulriche's International Periodical Directory* is a useful reference to know about because it lists all periodicals and the sources in which they are indexed.

One major directory of abstracts that is not frequently used but that is extremely helpful in locating sources is *Dissertation Abstracts International.* Because a literature review in a dissertation is usually exhaustive, obtaining a dissertation in your topic area may provide the most comprehensive literature review and bibliography available.

Searching Books and Other Documents

Most libraries, particularly those in colleges and universities, have already computerized the listing of their collections of books and documents. As in searching periodicals, on-line book searching is equally as efficient and fun. In these searches, you obtain information about the title, author, and source, as well as about the subjects, listed by key words and concepts. Also a description of the location of the book and its availability are included in the full citation. If you are unfamiliar with conducting a search, ask the librarian how to use the library system.

Most beginning investigators are concerned with how many sources to review. Although there is no magical number, most researchers search the literature for articles written within the previous 5 years. A search is extended back in time if necessary to evaluate the historical development of an issue or to review perhaps a breakthrough article that was published earlier. The depth of a review is dependent on the purpose and scope of your study. Most investigators try to become familiar with the most important authors, studies, and papers in their topic area. You will have a sense that your review is comprehensive when you begin to see citations of authors you have already read. Also, do not hesitate to call a researcher who has published in your topic area and ask for his or her recent work or articles he or she would recommend you review.

Step 4: Organizing Information

With your list of sources identified by your searches, you are now ready to retrieve and organize them. We usually begin by reading abstracts of journal articles to determine their value to the study. Based on the abstract, we determine if the study fits one of four categories as suggested by Findley.[6] If a study is highly relevant, it goes into what Findley calls the A pile: those studies that absolutely must be read. If the study appears somewhat relevant, it goes into a B pile: those studies that we probably will read, depending on the direction we take. If the study might be relevant, it goes into a C pile: those studies we might read, depending on the direction we take. Finally, if a study is not relevant at all, it goes into an X pile. These are studies that we do not want to read but keep just in case.

With books we do the same type of categorizing based on the table of contents and the preface.

Next, we start with the A pile and begin to classify the articles according to the major subjects or concepts. It is helpful to pick a reading order to outline a preliminary conceptual framework for your literature review.

Because it is not possible to remember all the important information presented in reading, taking notes is critical. There are many different schemes for note taking. Some find that taking notes on index cards is effective in that the index cards can be shuffled and reordered to fit into an outline. Others take notes on a word processing system. In general, your notes should include a synopsis of the content, the conceptual framework for the work, the method, and a brief review of the findings. Whatever method you use, be sure to always cite the full reference. All researchers have had the miserable experience of losing the volume or page number of a source and then having to spend hours finding it later on.

TABLE 5–1.

Literature Review Chart

Author (Date)	Sample Boundaries	Design	Dependent Constructs	Claims
Blumenthal et al.[8] (1982)	24 elderly from retirement homes (mean age 69 yr)	No control group	Mood (profile of mood)	No change
			Temperament (Thurstone scale)	No change
			Self-report of improvement	40% improved
			Personality type	No change

Although each investigator organizes and writes in his or her unique way, we offer two strategies you may want to try or modify to fit your own purpose and style. We suggest using a chart and concept/construct matrix to organize your reading. These two organizational aids will also assist in the writing stage of your review.

Charting the Literature

In this approach, you select pertinent information from each article you review and record it on a chart. You first need to determine the categories of information you will record or place on the chart. The categories you choose should reflect the nature of your research project and how you plan to develop a rationale for your study. For example, if you are planning an experimental exercise study for the elderly with diabetes, your review might highlight the limited number of adequately designed exercise studies with this particular population. The categories of your chart might, therefore, include the designs of previous studies and health status of the samples. An excerpt of a chart is shown in Table 5–1 to illustrate this method of organization. This chart was developed for a review of literature on exercise intervention studies for healthy older adults.

Charting research articles in this way will allow you to reflect on the literature as a whole and critically evaluate and identify research gaps.

Concept/Construct Matrix

The concept/construct matrix organizes information you have reviewed and evaluated by key concept/construct (the X axis) and source (the Y axis). Table 5–2

TABLE 5–2.

Butler's Concept/Construct Matrix

	Concept 1	Concept 2	Concept 3	Concept 4	Concept 5
Concept Source	Homelessness Hosp and Com 1100–1100	Gender Miller	Middle age Hunter and Sundrel	Pathology DSMIII-R	Poverty Axinn and Levin

displays an excerpt from a concept matrix used in a study by Butler.[7] In this naturalistic study of homeless, middle-aged women, Butler began her literature review by searching three major constructs: homelessness, gender, and middle age. Two other constructs, pathology and poverty, were also identified in the literature on homeless persons. Butler integrated these constructs into her theoretical discussion of existing knowledge, and they were added to the concept matrix.

As you begin to write the literature review, the concept/construct matrix facilitates quick identification of the articles that address the specific concepts/constructs you need to discuss.

Step 5: Critically Evaluating the Literature

There are several guiding questions you can use to help you critically evaluate the literature. Use the questions in Table 5–3 to guide your reading of

TABLE 5–3.

Questions for Analysis of Research

1. Was the study clear, unambiguous, and internally consistent?
2. What is (are) the research query(s), or question? Are they clearly and adequately stated?
3. What is the purpose of the study? How does the purpose influence the design and the conclusions?
4. Describe the theory that guides the study and the conceptual framework for the project. Are they clearly presented and relevant to the study?
5. What are the key constructs identified in the literature review?
6. What level of theory is suggested in the literature review? Is it consistent with the selected research strategy?
7. What is the rationale for the design found in the literature review? Is it sound? Does the design of the project fit the level of theory, and is relevant knowledge presented in the literature review?
8. Diagram or describe the design.
9. Does the design answer the research query(s)? Why or why not?
10. What are the boundaries of the study?
11. How are the boundaries selected?
12. What efforts did the investigator make to ensure trustworthiness or validity and reliability?
13. What data collection techniques were used? Is the rationale for these techniques specified in the literature review and/or in the methods section? How does data collection fit with the study purpose and study question, or query?
14. How are the data analyzed? Does the analysis plan make sense for the study? How does the analysis plan fit with the study purpose and study question or query?
15. Are the conclusions supported by the study?
16. What are the limitations of the study? What are the strengths of the study?
17. What level of knowledge is generated?
18. What use does this knowledge have for health and human service practice?
19. Are there ethical dilemmas presented in this article? What are they? Did the author(s) resolve the dilemmas in a reasonable and ethical manner?

TABLE 5–4.

Guiding Questions for Evaluating Non-research Sources

1. What way of knowing and level of knowledge are presented?
2. Was the work presented clearly, unambiguously, and consistently?
3. What is the purpose of the work? Implicit? Stated? How does the purpose influence the knowledge discussed in the work?
4. What is the scope and/or application of the paper?
5. What support exists for the claims being made in the source?
6. What debates, new ideas, and trends are presented in the work?
7. What are the strengths and weaknesses of the work?
8. What research queries, or questions, emerge from the work?

research literature and the questions in Table 5–4 to prompt your evaluation of non-research literature. Your responses to these evaluative questions will inform your research direction.

An evaluation of non-research sources involves examining the level of and support for the knowledge presented. Non-research sources may include position papers, theoretical works, editorials, and debates. Another important resource is work that presents a critical review of research in a particular area. These published literature reviews provide a synthesis of research as a way of summarizing the state of knowledge in a particular field and identifying new directions for research.

Step 6: Writing the Literature Review

You have now searched, obtained, read, and organized your literature. Finally, it is time to write a review. The review is written for a proposal to justify doing the research or for a manuscript that describes the completed research project.

A good literature review not only presents an overview of the relevant work on your topic but also includes a *critical* evaluation of the works. There is no recipe for writing the narrative literature review. Most investigators, regardless of where they fall on the continua of designs, include the following categories of information in their narratives:

- Introduction
- Discussion of each related concept, construct, principle, theory, and/or model in the current literature
- Brief review of related study designs and results
- Critical appraisal of current related research and/or knowledge
- Integration of the various works reviewed

- Niche that the investigator's study will fill within the collective knowledge related to the topic under investigation
- Overview and justification for the study and design

A suggested outline for writing the narrative follows:

Outline for Writing a Review

    I. Introduction (overview of what the review covers)
   II. Review of specific concepts
      A. How each concept has been studied
         1. Overview of studies
           a. Design
           b. Results
           c. Critical evaluation
         2. Critical evaluation of current knowledge
  III. Integration of concepts
      A. Relationships proposed in studies
      B. Identification of gaps in the literature
      C. Identification of research needs
  IV. Rationale for study and design
   V. Overview and niche for the study

SUMMARY

In this chapter we have discussed four major purposes of a literature review and six steps in conducting a search. We also have provided guidelines for writing an effective review.

The literature review is a critical evaluation of existing literature that is relevant to your study. Because it is a critical evaluation, reviewing the literature is initially a thinking process that involves piecing together and integrating diverse bodies of literature. The review provides an understanding of the level of theory and knowledge development that exists about your topic. This understanding is essential so that you can determine how your study fits into the construction of knowledge in the topic area.

The literature review is also an action process whereby you organize concepts and logically present your sources in a written review. The literature review chart and concept/construct matrix are two organizational tools to help you conceptualize the literature and write an effective review that convinces the reader that your study is necessary and is the next step in knowledge building. Remember that the primary purpose of writing a literature review and including it in the written report of your research is to establish the conceptual foundation for your research question, to establish the specific content of your study, and, most important, to provide a rationale for your research design.

EXERCISES

1. You want to study the effectiveness of a joint mobilization program to increase the range of motion of the lower extremities of children with spinal cord injuries. Define the key words you would use to conduct a search for relevant literature.

2. Select a research article and develop a concept/construct matrix from the literature presented. What level of knowledge and theory development is presented? Was the design selected appropriate for the level of knowledge in the literature?

REFERENCES

1. Kaplan S: Franco-American culture: knowledge by service providers, thesis, 1992, Northampton, Mass, Smith School for Social Work.
2. DePoy E: Mastery in clinical occupational therapy, *Am J Occup Ther* 44:55–59, 1991.
3. Bailey DM: Reasons for attrition from occupational therapy, *Am J Occup Ther* 4:23–29, 1990.
4. Willoughby J, Keating N: Being in control: the process of caring for a relative with Alzheimer's disease, *Qualitative Health Res* 1:27–50, 1991.
5. Maruyama M: Endogenous research: the prison project. In Reason P, Rowan J, editors: *Human inquiry: a sourcebook for new paradigm research*, New York, 1981, John Wiley & Sons, pp 267–282.
6. Findley TM: The conceptual review of the literature or how to read more articles than you ever wanted to see in your entire LIFE, *Am J Phys Med Rehabil* 68:97–102, 1989.
7. Butler S: Perspectives on the lives and needs of homeless, middle-aged women, unpublished dissertation, 1992, Seattle, University of Washington.
8. Blumenthal JA, Schocker DD, Needels TL, et al: Psychological and physiological effects of physical conditioning on the elderly. *J Psychosom Res* 26:505–510, 1982.

6

Design Classification

- **Classifying Designs by Purpose**
- **Classifying Designs by Structure**
- **Classifying Designs by Element of Time**
- **Classifying Designs by Content**
- **Classifying Designs by Context**
- **Summary**

Now that you have developed either a line of inquiry or a precise research question and have grounded the research problem in the literature, you are ready to select a design.

Selecting a specific design and outlining the methods and procedures are two of the most demanding efforts of the research process. They involve both creativity and knowledge of design strategies as you apply research principles to actual research situations. There are many factors to consider in developing a design. In the chapters that follow, we shall discuss specific design principles for both experimental-type research and naturalistic inquiry.

Before we begin to examine each design continuum in the chapters that follow, we would like to introduce you to the ways in which other researchers and texts classify design structures. Knowing how others organize and categorize designs will help you become a better consumer and evaluator of research.

There are numerous classification systems for research designs. Here we discuss five basic ways by which designs have been classified: (1) purpose, (2) structure, (3) element of time, (4) content and, (5) context. Let us examine each.

CLASSIFYING DESIGNS BY PURPOSE

Designs are often classified according to their purpose or by the type of knowledge that will be produced. The most widely used scheme is the classification of designs in three levels: that of exploratory, descriptive, and explanatory research.[1, 2]

1. *Exploratory research.* Included in the exploratory category are studies conducted in natural settings with the explicit purpose of discovering phenomena, variables, theory, or combination thereof.[3] Exploratory studies are widely varied in their structures and are characterized by no active manipulation or alterations of the research context or its conditions by the researcher. Studies in this category may be done in field and natural settings or in non-natural settings. Exploratory research may include both inductive designs that are capable of discovery and theory generation and deductive designs that examine the characteristics of specified variables.[4] An example of an exploratory study from an inductive perspective is *Tally's Corner* by Liebow.[5] Through the use of extensive observation and interview techniques (discussed in Chapter 15), Liebow obtained stories of men's lives on Tally's Corner, an urban street corner in a black neighborhood, and further was able to explore the meanings of their experiences. Liebow wanted to illuminate the richness of the lives of these men, as well as struggles that each man encountered when attempting to fill varied life roles. His purpose in conducting an exploratory study was to reveal theory, not to support or test existing theory.

2. *Descriptive research.* The purpose of descriptive studies is to yield descriptive knowledge of population parameters and relationships among those parameters. This category of research is anchored on a priori theory. It is a category of research that is frequently used in health and human services. Structurally, descriptive projects have preconceptualized areas of interest to which the study is limited and that are well defined by the a priori theory. These studies tend to rely heavily on measurement and quantitative analytic methods.

For example, let us say you would like to find out about the specific population characteristics of those with spinal cord injury. This information would be helpful to plan health services and specific intervention approaches for patients and their family members. This level of description involves identification of the frequency of occurrence of factors determined to be important by the investigator, such as gender, age, ethnicity, and religion. Perhaps it would also be important to obtain information regarding the extent to which young adults with spinal cord injury who were drug abusers also experience depression. This level of description involves the identification of relationships between two variables, that of drug abuse and depression. In examining a relationship or correlation between variables, the researcher examines the way in which the two factors relate to each other or how variations in scores on one variable are related to variations in scores on the other variable. It is important to keep in mind that a descriptive study provides knowledge about relationships but does not show that one variable causes another.

Let us look at an example to clarify this important point. Suppose you were to conduct a descriptive study that examined the relationship between chocolate ice cream consumption and schizophrenia in young adults. The results of your study showed that a disproportionately high number of persons with schizophrenia consumed a greater amount of chocolate ice cream when compared with a group of persons who did not have schizophrenia. In other words, a strong relationship was revealed between schizophrenia and chocolate ice cream consumption. Based on this relationship, can you claim that eating chocolate ice cream caused schizophrenia? We hope your answer was no, but you might want to investigate the relationship further with an explanatory design.

3. *Explanatory research.* Explanatory studies, also called experimental research,[6] are designed to reveal causal relationships and predict outcomes as well. This category of research design is considered the most desirable and powerful design in the positivist world. Explanatory studies are founded on an accepted theoretical frame of reference and seek to support theory through hypothesis testing and prediction. To show causal links, studies are structured to eliminate the effects of extraneous influences on the outcome or dependent variable that is being measured. Explanatory studies rely only on measurement and statistical analysis of quantitative data.

Let us say, for example, that you wanted to know if a hospital-based group socialization program was effective in reducing depression in young adults with a spinal cord injury. You might structure the study so that these patients were randomly assigned to either an experimental group, in which the socialization program was conducted, or to a control group, which received individual counseling. You would use a standardized depression scale that fit your theoretical framework. All participants would be tested on the scale before their participation in the experimental and control group programs and after the conclusion of the programs. A statistical comparison of the changes in depression scores between the two groups would be used to determine the extent to which there is a causal link between your program and lowered depression.

There are two important limitations to classifying designs according to three levels of purpose. First, both naturalistic and deductive-type studies are often linked together at the exploratory level of purpose. Thus, the purposive classification system overlooks the different perspectives of each and does not distinguish among the diverse purposes and approaches of designs on the naturalistic continuum. Second, the classification assumes that naturalistic inquiry can be used only at the exploratory level. Many qualitative researchers would argue that their research purpose and the knowledge generated by their investigations provide rich description and explanation of relationships.[7, 8] This classification system is effective, however, in understanding the three levels of purpose for the experimental-type continuum. In our system of classifying designs, we refer to these three levels of purpose to examine experimental-type research only.

CLASSIFYING DESIGNS BY STRUCTURE

Designs have also been classified by the way in which data gathering and analysis is organized and structured. There are two basic ways in which designs have been classified by structure.

The first classification system is used to examine and analyze the structure of experimental-type designs. Developed primarily by Campbell and Stanley,[6] in this scheme there are three levels of research: non-experimental, quasi-experimental, and experimental. As you can see by the nomenclature, the classification system excludes all research that is not based on the assumptions of the experimental paradigm. Designs are recognized as either those that are experimental, those that are somewhat like experimental but are missing something (quasi-experimental) or those that are missing (non-experimental) the essential elements of experimental design. Essentially, Campbell and Stanley analyzed explanatory and descriptive levels of research in detail but did not fully attend to exploratory designs. Campbell and Stanley characterized each design by specific attributes and diagrammed designs by using the symbols of **X** (manipulated variable) and **O** (observation points) to represent the sequence and activities of an investigation. Because we find this system an excellent aid for diagramming and understanding the experimental-type continuum, we present it in detail in that chapter.

The second structural approach to classification is based on the nature of the data collected and the type of analysis conducted. In this scheme, studies are classified as either "qualitative" or "quantitative" in their structure. As we have indicated in previous chapters, the qualitative-quantitative dichotomous classification separates studies into methodologic opposites: those that use numerical data and statistical analysis and those that rely on narrative and qualitative-type analyses. Although it may be easy to classify studies in this way, it is very misleading and simplistic. This approach does not adequately portray the structural differences of designs or reflect the rich differences that exist among researchers conducting naturalistic inquiry. Furthermore, classifying designs as polar opposites does not provide a basis from which designs from either side can be integrated and used together to enhance the other.

CLASSIFYING DESIGNS BY ELEMENT OF TIME

Designs are also classified by the element of time. "Time" is an important factor in the design and execution of research along both the experimental-type and naturalistic continua. There are two basic design classifications along the dimension of time: retrospective and prospective research. Let us examine each classification.

Retrospective Studies

As the name implies, retrospective studies describe and examine phenomena after the fact or after the phenomena have occurred. In retrospective studies, the effect or outcome is already known, but the cause or nature of phenomena is not understood. Along the experimental-type continuum, examples of retrospective designs include those that use chart extraction and are classified as passive observation or correlational. Although there is no manipulation of a phenomenon or random assignment or control, the purposes of such studies may vary and include (1) description of the occurrence of a phenomenon, (2) examination of relationships among variables, and (3) examination of possible causative relationships.

When a chart review is conducted or information is extracted from medical records, there are two critical issues. One concerns the reliability or consistency of the initial data recordings, and the second issue concerns the accuracy in the way in which information is extracted from the chart. The researcher must establish procedures to check for initial reliability of chart information and inter-rater reliability or accuracy of coders.

A retrospective approach is useful when a reasonable data set, such as medical records, is available, if the researcher has limited time and funding resources, if it is not feasible to examine cause-effect relationships as they occur, or if random assignment is not appropriate. Davidoff et al.[9] conducted a retrospective study to determine the efficacy of rehabilitation programs to facilitate recovery after acute stroke. They reviewed the medical records of 139 acute stroke patients at admission, discharge, and 1-year follow-up from hospitalization. They examined such information as neurologic status, functional status, and the quantity and reason of outpatient physical and occupational therapy. Through the retrospective analysis of events that had already occurred, the authors were able to derive two very important findings. First, they examined functional gains from onset to discharge for all patients and found that all patients made substantial gains in self-care and mobility skills during their rehabilitation course. Second, they divided the subject pool into two groups: those who received outpatient therapies and those who did not. They then compared the functional gains of both and determined the factors associated with receiving outpatient therapies. They found that although the outpatient therapy group demonstrated inferior functional performance at each repeated measure compared with the no therapy group, they did show improvement in mobility and self-care skills. This is an important finding in that it suggests that further functional gains appear to be associated with additional outpatient therapies for some patients.

The researchers in this study were unable to manipulate the independent variable, in this case the receipt of rehabilitation and outpatient therapy, to examine their effectiveness. However, this correlational design using three post hoc measurement points (admission, discharge, and 1-year after rehabilitation) provides information as to the relationship between rehabilitation and the maintenance of gains. In this way the study adds incrementally to the body of literature that is attempting to demonstrate the effectiveness of rehabilitation and outpatient therapy on the long-term functional gains of individuals with stroke.

Along the naturalistic design continuum, the life history approach is an example of a retrospective perspective. In this approach, the investigator works with the informant to recall and document the major life experiences and events as a way of understanding that individual. In "qualitative" research with the elderly, another useful interview technique is that of "reminiscence." In this type of in-depth interviewing, the interviewer works with the informant to go back in time and reflect on personal experiences and feelings and to "reminisce." The rethinking of the past produces narrative representing an individual's recollections and interpretations. This retelling and recasting of life events, feelings, and experiences can be used as a basis from which to interpret how individuals make sense of "self" as they age.

Prospective Studies

In a prospective approach, the purpose is to describe phenomena, search for cause and effect relationships, or examine change in the present or as the event unfolds over time. There are two basic types of prospective designs: cross-sectional and longitudinal. These prospective approaches to design can be used with a range of data collection strategies that vary in their degree of structure.

Cross-sectional Study

In a cross-sectional study design the researcher examines a phenomenon at *one* point in time. That is, there is only one single time frame in which data are collected. Suppose you wanted to describe attitudes toward aging and how individuals of varying ages perceive the aging process. You would design a cross-sectional study that would involve obtaining a sample of individuals in different age groups. You would then either conduct a survey or interview at one point in time. The analysis would involve a comparison among the age groups on the dependent variables, such as attitudes toward aging. For example, all interviews would occur in the same time frame, let us say, in 1990, with individuals representing each age group.

1990
Age group (yr)
41–50
51–60
60–70

Longitudinal Studies

Longitudinal studies involve data collection over extended periods of time. These are important but costly approaches to examine such phenomena as the long-term effects of health care programs or interventions, the natural course of human development and adaptation, the trajectory of illness, or the sequelae of various diagnoses. Three types of longitudinal studies are described: trend, cohort, and panel studies.

Trend studies involve examining a general population over time to see changes or trends that emerge as a consequence of time. Let us use the example just discussed. Suppose you wanted to see if there were any trends in the way in which different age groups perceived the aging process over time. You would interview individuals who fit into the age brackets you were interested in; for example:

1990–1995	
Age Group (yr)	
1990	1995
41–50	41–50
51–60	51–60
61–70	61–70

These individuals would be interviewed at one point in time as in the cross-sectional design. Then 5 years later, let us say, in 1995, you would interview other individuals who fit the same age brackets. In this way you would be able to compare and contrast attitudes toward aging of persons 41 to 50 years old in 1990 with persons of that same age range in 1995. You could also repeat the interview with different individuals 5 years later in the year 2000 to see if there were any trends or changes that occurred as a consequence of time.

Cohort studies involve examining a specific group or one particular generational grouping as they change over time. A cohort, or one particular generational grouping, would be followed over time. For example, in the study on attitudes toward aging, we would examine aging perceptions of one age group at one point in time, such as 1990. Then, at a later point in time, let us say, 1995, we would sample different individuals but those who are from the same cohort as in the first data collection effort. The individuals in the sample would thus be 5 years older. Although we would be sampling from the same cohort of individuals, we would not be interviewing the same individuals at both time periods.

Comparisons of Different Subjects at Each Time or Testing Occasion

Age Group (yr)	
1990	1995
41–50	46–55
51–60	56–65
61–70	66–75

Panel studies represent another important longitudinal design strategy. It is similar to cohort design except the *same* set of people are studied over time. For example, the subjects we interviewed in 1990 who were 41 to 50 years of age would be the same individuals we would attempt to interview 5 years later in 1995.

Comparison of Same Subjects at Each Time or Testing Occasion

Age Group (yr)	
1990	1995
41–50	46–55
51–60	56–65
61–70	66–75
71–80	76–85

Sometimes a longitudinal panel design is used in combination with a true experimental type design to examine the long-term effects of an intervention.

CLASSIFYING DESIGNS BY CONTENT

Designs may also be classified by the content that the researcher is investigating. For example, social science research may be categorized by its focus on a particular stage in the life span such as gerontologic research or childhood developmental research. Specific design strategies have emerged to address the issues of conducting research with specific populations. For example, in the field of aging there are twists and variations in designs in naturalistic inquiry that are used when older populations are studied. Along the experimental-type continuum, specific sequential crossover designs have been developed to determine if observations are the result of age-related dynamics or other biopsychosocial phenomena.

Research may also be classified by substantive content in a particular discipline. For example, many texts are written specifically for one discipline.

David Currier[10] wrote a text specifically for research in physical therapy, Reid and Smith[11] authored *Research in Social Work*, Burns and Grove[12] wrote a text exclusively for nursing research, and Bailey[13] wrote a text for occupational therapists. In addition to the standard designs presented in any research text, each of these texts also present specific design strategies that are directly relevant to the discipline and offer examples of disciplinary-specific research.

CLASSIFYING DESIGNS BY CONTEXT

Studies are also classified by the context in which they are conducted. There are three major categories of studies in this classification approach: field, laboratory, and clinical studies.[2] Field studies refer to research conducted in natural settings, whereas laboratory studies are implemented in controlled environments. In health care research, clinical research is a third category, which refers to the investigation of human experience within the context of health care institutions. Clinical research may be based in the field or laboratory, depending on the nature of the research query. For example, a study of the culture of a nursing home is in essence a study of a clinical setting. However, because the study seeks to reveal the culture in the natural setting of the institution, it is a field study. On the other hand, a study of the responses of nursing home residents to a new activity program may be considered clinical research. But it is also a laboratory setting in that the nursing home environment, in this case, is a controlled setting where events are being manipulated by the researcher. Classifying designs by context facilitates an understanding of how the setting establishes the parameters for a design structure.

SUMMARY

We have presented five major ways designs are discussed and classified. You will see references to each of these classification schemes in your readings of other research texts and articles. Although each scheme provides some understanding of the characteristics of designs, we have also pointed out their limitations. As indicated in Chapter 2, rather than use any one of these classification systems, we organize designs along two design continua. This model provides the organization and criteria on which to judge the merits and limitations of designs based on their own philosophical and theoretical perspective and basic purpose. Within the continua of design classifications for experimental-type and naturalistic inquiry, we borrow aspects from each of the five schemes presented in this chapter when it is appropriate and useful to do so.

REFERENCES

1. Adams GR, Schvansveldt JD: *Understanding research methods,* ed 2, New York, 1991, Longman.
2. Tripodi T, Fellin P, Meyer H: *The assessment of social research,* Itasca, Ill, 1983, Peacock.
3. Glaser BG, Strauss AL: *The discovery of grounded theory: strategies for qualitative research,* New York, 1967, Aldine.
4. Reid WJ, Smith AD: *Research in social work,* ed 2, New York, 1989, Columbia University Press.
5. Liebow E: *Tally's corner,* Boston, 1967, Little, Brown.
6. Campbell DT, Stanley JC: *Experimental and quasi-experimental design,* Chicago, 1963, Rand McNally.
7. Agar MH: *Thinking of ethnography,* Newbury Park, Calif., 1986, Sage.
8. Spradley J: *Participant observation,* New York, 1980, Holt, Rinehart & Winston.
9. Davidoff GN, Keren O, Ring H, et al: Acute stroke patients: long term effects of rehabilitation and maintenance of gains, *Arch Phys Med Rehabil* 72:869–873, 1991.
10. Currier DP: *Elements of research in physical therapy,* Baltimore, 1984, Williams & Wilkins.
11. Reid WJ, Smith AD: *Research in social work practice,* New York, 1989, Columbia University Press.
12. Burns N, Grove S: *The practice of nursing research: conduct, critique and utilization,* Philadelphia, 1987, FA Davis.
13. Bailey D: *Research for the health professional: a practical guide,* Philadelphia, 1991, FA Davis.

7 | Experimental-Type Designs: Language and Thinking Processes

- **Defining Design in Experimental-Type Research Tradition**
- **Sequence of Experimental-Type Research**
- **Structure of Experimental-Type Research**
- **Plan of Design**
- **Summary**
- **Exercises**

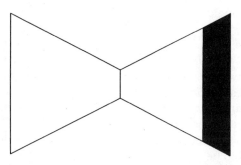

LANGUAGE AND THINKING PROCESSES
• Structure of design
• Plan of design

We are ready to explore in-depth the language and thinking of researchers who use the continuum of experimental-type designs. But let us first summarize what we have said in previous chapters about this continuum. Experimental-type designs are characterized by thinking and action processes based in deductive logic and a positivist paradigm. In this tradition, the researcher seeks to identify a single reality through systematized observation. This reality is understood by examining its parts and the relationship among these parts. Ultimately the purpose of this research is to predict what will occur in one part of the universe by knowing and observing the other part. Research along this continuum can be characterized as follows:

1. It is based in one epistemology.
2. It has been accepted historically as the "scientific method" for discovering "fact."
3. It has been evaluated and described by a unified and agreed on vocabulary that is well established.
4. There is strong consensus about which action processes are adequate and scientifically rigorous.
5. All designs on the continuum share a common language and a unified perspective as to what constitutes adequate design.

In this chapter we discuss the language and basic thinking processes used by researchers in developing experimental-type designs.

DEFINING DESIGN IN EXPERIMENTAL-TYPE RESEARCH TRADITION

There are many definitions of research design, each reflecting a different epistemologic perspective. However, in the experimental-type continuum, the term research design has a single and agreed on meaning. Research design is the *plan,* or *blueprint,* that specifies and structures the action processes of collecting, analyzing, and reporting data to answer a research question. As Kerlinger[1] states, design is "the plan, structure and strategy of investigation conceived so as to obtain answers to research questions and to control variance" [p. 300]. In this definition, "plan" refers to the blueprint for action or the specific procedures used to obtain empirical evidence. "Structure" represents a more complex concept and refers to a model of the relationships among the variables of a study. That is, design is structured in such a way as to enable an examination of an hypothesized relationship between variables. The main purpose of design is to structure the study so that the researcher can determine the extent to which one phencmenon (the independent variable) is responsible for change in another (the dependent variable).

A formal understanding of design is that its purpose is to control variance or restrict or control extraneous influences on the study. By exerting such control, the researcher can state with a degree of statistical assuredness that study outcomes are a consequence of either the manipulation of the independent variable (as in the case of experimental design) or the consequence of that which was observed and analyzed (as in the case of non-experimental designs). In other words, the design provides a degree of assuredness that an investigator's observations are not haphazard or of random events but reflect a "true" objective reality. The researcher is thus concerned with developing the most optimal design that eliminates or controls what researchers refer to as "disturbance," "variance," "extraneous factors," or "situational contaminants." The design controls these disturbances or situational contaminants through the implementation of systematic procedures and data collection efforts discussed in subsequent chapters. The purpose of imposing control and restrictions on observations of natural phenomena is to assure that the relationships specified in the research question or questions can be identified, understood, and ultimately predicted.

To summarize, then, design is what separates research from the everyday type of observations and thinking and action processes in which each of us engages. Design instructs the investigator to "do this; don't do that." It provides a mechanism of control to assure that data are collected "objectively," in a uniform manner, with minimal investigator involvement or bias. The important point to remember is that the investigator remains separate from and uninvolved with the phenomena under study and that procedures and systematic data collection provide mechanisms to control and eliminate bias.

SEQUENCE OF EXPERIMENTAL-TYPE RESEARCH

Let us review the sequence of thought and actions followed by experimental-type researchers to understand the central role of design. As Figure 7–1 illustrates, design is pivotal in the sequence. It stems from the thinking processes of formulating a problem statement, literature review, theory specific research question, and hypotheses or expected outcomes. In turn, the design dictates the nature of the action processes of data collection, the conditions under which observations will be made, and most important, the type of data analyses and reporting that will be possible.

Recall our previous discussions in which we illustrated how the problem statement identifies the purpose of the research and the broad topic an investigator wishes to address. The literature review then guides the selection of theoretical principles and concepts of the study and provides the rationale for a research project. This rationale is based on the nature of research previously conducted and the level of theory development for the phenomena under investigation. The experimental-type researcher must develop a literature review in which the specific theory and measures that he or she will use in a study are discussed and supported as "the best." Furthermore, the type of research question asked must be consistent with the knowledge reported in the literature.

Thus, the researcher develops a design that builds on the ideas and actions that have been conducted and in ways that conform to the rules of scientific rigor. One's choice of a design not only is shaped by the literature and level of theory development but also depends on the specific question that is asked. There is nothing inherently good or bad about a design. Every research study and

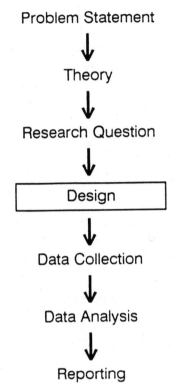

FIG 7–1.
Sequence of experimental-type research.

design has its own particular strengths and weaknesses. The adequacy of a design is based on how well the design can answer the question that is posed.

STRUCTURE OF EXPERIMENTAL-TYPE RESEARCH

We have said that experimental-type research has a well-developed language that sets clear rules and expectations for the adequacy of design and research procedures. Table 7–1 summarizes the meaning of nine key terms used by researchers to structure designs along the continuum of experimental-type research.

Concept

We defined this term in our discussion of theory in Chapter 3 as the words or ideas that symbolically represent observations and experiences. Concepts in and of themselves are not directly observable. Rather, what they describe can be observed or experienced. Concepts are "(1) tentative, (2) based on agreement and (3) useful only to the degree that they capture or isolate something significant and definable."[2] For example, the terms "grooming" and "work" are both concepts that describe specific observable or experienced activities people engage in on a regular basis. Other concepts such as "personal hygiene" or "sadness" have various definitions, each of which have led to the development of assessment instruments to measure the underlying concept.

Constructs

Constructs are defined by Babbie as "theoretical creations based on observations but which cannot be observed directly or indirectly."[3] Constructs can only be inferred and may represent a model of a relationship between two or more concepts. For example, the construct of "community function" may be inferred from observing concepts such as "grooming," "personal hygiene," and "work." However, the construct is not directly observable unless it is broken down into its component parts or concepts.

Definitions

There are two basic types of definitions that are relevant to research design: conceptual definitions and operational definitions. *Conceptual definitions* are those that stipulate the meaning of concepts or constructs with other concepts or constructs. Operational definitions are those that stipulate meaning by observation or experience. *Operational definitions* "define things by what they do." For example, the construct of self-care may be conceptually defined as the activities that are necessary to care for one's bodily functions, whereas it is

TABLE 7–1.

Key Terms in Structuring Experimental-Type Research

Term	Definition
Concept	Use of words to symbolically represent observation and experience
Construct	Represents a model of relationships between two or more concepts
Conceptual definitions	The meaning of a concept expressed in words
Operational definitions	How the concept will be measured
Variable	An operational definition of a concept assigned numeric values
Independent variable	The presumed cause of the dependent variable
Intervening variable	Those phenomena that have an effect on study variables
Dependent variable	The variable that is affected by the independent variable or is the presumed effect or outcome
Hypotheses	Testable statements that indicate what the researcher expects to find

operationally defined as observations of an individual engaging in bathing, dressing, and grooming.

Variables

A variable is a concept or construct to which numeric values are assigned. By definition, a variable must have more than one value even if the investigator is interested in only one condition. For example, if an investigator was interested in determining the self-care routine of persons with disabilities who were employed, both "self-care routine" and "employment" would be variables. However, although self-care routine may have multiple values, the investigator undertaking this inquiry is interested in only one of the potential values that could be attributed to employment status, that of being employed.

There are three basic types of variables: independent, intervening, and dependent. "*An independent variable* is the presumed cause of the *dependent variable*, the presumed effect."[4] That is, the independent variable most always precedes the dependent variable and has a potential influence on it. It is also sometimes referred to as the predictor variable. *Intervening variables*, also called *confounding* or *extraneous variables*, are those phenomena that have an effect on the study variables but are not necessarily the object of the study. In the study just mentioned about self-care routine of employed individuals with disability, the variable "employment status" represents the independent variable, whereas "self-care routine" represents the dependent variable. Intervening variables in this study may include any other variable that potentially influences either of the independent or dependent variable, such as family support, degree of disability, motivational status, and functional ability. *Dependent variables*, also called *outcome* and *criterion*, refer to that which the investigator seeks to understand, explain, or predict.

Hypotheses

Hypotheses are defined as testable statements that indicate what the researcher expects to find based on theory and level of knowledge in the literature. Hypotheses are stated in such a way so that they can either be verified or refuted by the research process. The researcher can develop either a directional or nondirectional hypothesis. In a directional hypothesis, the researcher indicates whether she or he expects to find a positive relationship or an inverse relationship between two or more variables. In a positive relationship, the researcher may hypothesize that employed individuals with disabilities have greater self-esteem than unemployed individuals with disabilities. An inverse relationship would involve a hypothesis something like: Individuals with disability who remain employed will demonstrate less difficulty in performing self-care routines than those who become unemployed.

Let us summarize how these nine key terms are used. Research questions narrow the scope of the inquiry to specific concepts, constructs, or both. These concepts are then defined through the literature and operationalized into variables that will be investigated descriptively, relationally, or predictively. The hypothesis establishes an equation by which independent and dependent variables are examined and tested.

PLAN OF DESIGN

The plan of a design requires a set of thinking processes in which five core issues — bias, manipulation, control, validity, and reliability — are considered by the experimental-type researcher.

Bias

Bias is defined as the potential unintended or unavoidable effect on study outcomes. Many factors can cause bias in a study, such as:

- Selection of inappropriate instrumentation
- Sample selection process that favors a particular unintended group
- Improper training of interviewers
- Deviation from the plan and structure of the design

Let us consider each source of bias. There are two major issues in instrumentation: bias resulting from inappropriate data collection procedures and bias resulting from inadequate questions. Suppose you are a counselor working in a community mental health center and your boss, the agency administrator, is conducting a survey of worker satisfaction. To obtain information from employees, the boss decides to interview people. Knowing that this person is your superior, how likely are you to provide responses to job satisfaction questions, particularly knowing that your performance evaluation is next month? Using the same example, suppose a question asked is "Don't you agree that this is a great place to work?" Certainly this question is very poorly phrased in that it implies a socially and politically correct response. Interview questions that are posed in a way that elicit a socially correct response or questions that are vague, unclear, or ambiguous introduce a source of bias into the study design.

Now suppose that only the boss's social friends were selected as a sample to participate in the survey and are claimed to represent the agency population. Such a sample obviously would not represent the overall level of job satisfaction for all employees. The characteristics of the sample serve as potential bias or influence on the study's outcomes.

If, in this same study, multiple persons collected information but each asked questions differently about job satisfaction, interviewer procedures would serve as a strong bias. Improper or uneven training of data collectors can bias the outcomes of a study.

Experimental-type designs are characterized by the elaboration of plans and procedures before the conduct of the study. Thus, any deviation from the original plan for data collection and analysis may introduce bias. For example, suppose that the agency director decided to collect data regarding level of job satisfaction of the agency employees on Monday. Monday was selected because all employees were available. However, instead, the data collection was accomplished on Tuesday because Monday was a holiday. On Tuesday, all home-based service providers were in the field and therefore were not surveyed for their degree of job satisfaction. This deviation in the plan would significantly effect the study outcome in that a large segment of the agency employee population would have been omitted from the study.

There are many causes of bias in experimental-type design. An essential part of planning in design is to introduce systematic procedures to minimize or eliminate as many sources of bias as possible.

Manipulation

Manipulation is defined as the action process of maneuvering the independent variable so that the effect of its presence, absence, or degree can be observed on the dependent variable. To the extent possible, experimental-type researchers attempt to isolate the independent and dependent variables and then to manipulate or change conditions so that the cause and effect relationship between the variables can be examined. Goodman's[5] study of the effects of two types of support groups on the well-being and perceived social support of the caregivers of persons with Alzheimer's illustrates the concept of manipulation. To determine which of two interventions promoted positive outcome, Goodman tested the subjects to ascertain their well-being and perceived social support before any intervention. They were randomly assigned (assigned on a chance-determined basis) to each intervention and then tested again after the intervention. The subjects were then asked to switch groups and participate in the intervention that they had not experienced. They were tested again to determine if changes in well-being and perceived social support had occurred. In this example, the independent variables, or the interventions, were manipulated. By manipulating the circumstances under which a subject received the intervention, the investigator was able to examine the effect of each intervention and their combination on subject well-being and perceived social support. Goodman[5] was able to evaluate the independent effects and the joint effects of each intervention.

Control

Control may be defined as the set of action processes that direct or manipulate factors to achieve an outcome. Control plays a critical role in experimental-type design. Through controlling not only the independent variable but other aspects of the research context, such as subject assignment and the experimental environment, the relationship among the study variables can be most clearly observed. In experimental-type design, procedures to establish control are implemented to minimize the influence of extraneous variables on the outcome or dependent variable. In Goodman's[5] study, two basic methods of control were introduced: random group assignment and reversal of group assignment. By randomly assigning subjects to one group or the other, Goodman attempted to develop equivalence or eliminate subject bias due to inherent differences which may have occurred in the two groups. For example, let us say Goodman had not used random assignment but instead had just assigned the first ten volunteers to the first group and the next ten volunteers to the second group. It could be possible that the first ten subjects knew each other and entered the study first because they were highly motivated to seek help. On the contrary, suppose the last ten subjects were somewhat less motivated to participate. Familiarity with other subjects and high motivation to seek help, two factors inherent in the subjects assigned to group one, might influence the measures of well-being and perceived social support. By not using random assignment, the investigator would be at risk of developing groups that initially differed from each other. This initial difference then would make it difficult to discern what effect, if any, the independent variable might have on the outcome.

Another method to enhance control is the development of a control group. For example, Goodman might have randomly assigned subjects to a third group that did not receive any intervention. By comparing subject scores of this control group with the two intervention groups, Goodman would have been able to determine if changes in the dependent variable occurred just as a consequence of the passage of time, were the result of the attention factor inherent in participation in a study, or whether the particular content of each group had produced the change.

Validity

Validity is a concept that has numerous applications in experimental-type research. However, the concept as it applies to design refers to the extent to which one's findings are accurate or reflect the underlying purpose of the study. Although many classifications of validity have been developed, we discuss the four fundamental types of validity, based on Campbell and Stanley's[6] initial work: internal validity, external validity, statistical conclusion validity, and construct validity.

Internal Validity

Internal validity refers to the ability of the research design to accurately answer the research question. If a design has internal validity, the investigator can state with a degree of confidence that the reported outcomes are a consequence of the relationship between the independent variable and dependent variable and not the result of extraneous factors. Campbell and Stanley[6] have identified seven major factors that pose a threat to the ability of a researcher to determine whether what is being observed is a function of the study or due to external forces. These threats to internal validity include:

- History: the effect of external events on study outcomes.
- Testing: the effect of being observed or tested on outcomes.
- Instrumentation: the extent to which the instrument is accurate in its measurement and extent to which the instrument itself may be responsible for outcomes.
- Maturation: the effect of the passage of time.
- Regression: the effect of a statistical phenomenon in which extreme scores tend to regress or cluster around the mean (average) on repeated testing occasions.
- Mortality: the effect on outcome caused by subject attrition or dropping out of a study prior to its completion.
- Interactive effects: the extent to which each of these threats interacts with sample selection to influence the outcome of a study.

To illustrate these threats to validity, let us consider a hypothetical example. Suppose you are conducting a study to determine the extent to which an AIDS prevention program has reduced acquired immunodeficiency syndrome (AIDS) risk behavior in adolescents. Before the intervention, you test subjects to determine the extent to which they exhibit AIDS risk behavior. You then conduct a prevention program and retest the participants. The retest shows an amazing reduction in AIDS risk behavior, and you claim that your program is a success. In doing so, you are making a claim that a relationship exists between your program (the independent variable) and AIDS risk behavior (the dependent variable). Furthermore, you are claiming that there is a causal relationship between the variables. That is to say, you claim that your program caused the reduction in AIDS risk behavior.

Now consider how each of the seven threats to internal validity effect your research outcomes. First, let us say you have just learned that Magic Johnson held a news conference in which he announced that he is HIV positive. This announcement comes right after you first administered your test to your subjects but before the final testing occasion. It is certainly possible that this *historical event* rather than the independent variable was the factor responsible for reducing risk behavior in your subjects.

However, assume that your subjects already knew about Magic Johnson before beginning the study. Now consider the *threat of testing* to the validity of your claims. As a sensitive researcher, you have decided to measure AIDS risk behavior by observing a discussion group about sexual behaviors among subjects. Each time you hear a high-risk behavior, you score it as such without telling your subjects what you are doing. In the first testing situation, the adolescents appear to openly discuss their sexual decision making. But you also note that they are very aware of group norms and of being observed and stay well within the boundaries set tacitly by the group when discussing their personal behavior and views. In the second test, you note a large reduction in reported risk behaviors or in risky decision making expressed by the adolescents. Although it is possible that the prevention program is actually responsible for a risk reduction, there are also competing alternative explanations for the change. For example, it may be that in the first testing situation, adolescents revealed what was normative and acceptable in the group rather than accurately representing their risk potential. Or the observed reduction in risk behavior could conceivably be a function of the act of the testing situation itself. Now that participants are aware of what is being observed and recorded, they may answer more cautiously. The testing procedures pose yet another threat to the internal validity of the design. The reduction in reported risk behaviors may be a function of participants actively changing or adjusting their behaviors or thoughts on the topic as a consequence of participating in a group discussion with peers. The test itself or mode of data collection may have influenced a change in behavior independent of the effect of participating in the intervention. Therefore, it is difficult to determine whether change is a consequence of the test, a consequence of the intervention, or both.

This example also illustrates the threat to validity posed by *instrumentation.* In this example, the instrumentation may not be measuring what it was intended to assess. Instead of measuring risk behavior, the instrumentation may be a more accurate indicator of group norms related to sexual decision making.

Another way in which instrumentation poses a threat to a study is through changes that may occur within interviewers over time or with the instrument itself. For example, if you were collecting data regarding weight or blood pressure, any deviation in the calibration of a scale or blood pressure cuff from one testing occasion to the next would pose a significant threat to the validity of the data that are obtained. Or interviewer fatigue or any change in the way in which interviewers handle interview questions, for example, may cause deviation in responses that will effect the way in which the investigator can interpret the findings.

To illustrate the threat posed by *maturation,* let us assume that the AIDS prevention program just discussed is being held over a 1-year period. It is possible that the participants have matured in their thinking and have become more responsible in their decision making as a function of time alone rather than as a result of the prevention program.

The *threat of regression* frequently occurs when subjects with extreme scores are selected to participate in a study. In this example, suppose adolescents who are very sexually active participate in the study and therefore tend to report extremely high-risk behaviors. High scorers at the beginning of the program will tend to report less risky scores at the end. This change in scores, however, may be a consequence of a statistical principle known as "statistical regression toward the mean" in which extreme scores tend to move toward the mean on repeated testing occasions.

Now consider this scenario. Suppose 50 participants began the AIDS prevention program and only 10 remained by the final testing period. It is possible that as a result of *experimental mortality,* those who remained in the study were more committed to reducing risk behavior than those who dropped out. Thus, mortality or selective attrition of subjects poses a threat to the interpretation of study findings.

Numerous interactive effects could also confound the accuracy of the findings. For example, suppose that the sample that was selected for the study was a population of adolescents on probation. This group was highly motivated to report a change in their risk behavior (a sampling bias) and were also sensitive to being tested because of their judiciary status. The interactive effect of the sample bias and the testing condition may wreak havoc on claims that the outcome was a result of the prevention program.

External Validity

External validity refers to the capacity to generalize findings and develop inferences from the sample to the study population. External validity answers the question of generalizability. Let us use the example of the AIDS study to explore the issue of external validity. Assume that the investigator obtained a sample by asking for volunteers from a large group of adolescents between the ages of 12 and 18 years who were on probation in Philadelphia. The potential to generalize the findings from the sample to all adolescents on probation is limited by several important factors. First, the sample was a volunteer sample, making them potentially different from adolescents who did not volunteer. Second, the experiences of adolescents who live in large cities do not necessarily represent the experiences of adolescents in other geographic locations. Third, because the sample was a volunteer sample, the researcher cannot know in what ways the sample may differ from the larger population, which includes those who refused participation or did not have the opportunity to volunteer. Finally, the investigator limits generalizability to a very select population and will not be able to infer about the experiences of adolescents who, for example, are not on probation, live in rural communities, and may be sexually active and at risk but are not on probation. These factors represent just some of the threats to external validity that may occur in a study.

The internal validity and external validity of a study are very much related. As an investigator attempts to increase the internal validity of a study, the ability to make broad generalizations and inferences to a larger population decreases. That is, as more controls are implemented and extraneous factors are eliminated from a study design, the population to which findings can be generalized becomes more limited. Investigators always try to enhance the internal validity of a study to assure that valid and accurate conclusions can be drawn. Internal validity should be the investigator's primary concern, although it may limit the degree of external validity of one's findings.

Statistical Validity

Statistical validity refers to the power of one's study to draw statistical conclusions. One of the aims of experimental-type research is to find relationships among variables and ultimately to predict the nature and direction of these relationships. Support for relationships is contained in the action process of statistical analysis. However, as you will see when you read further in Chapters 8 and 17, selection of statistical techniques is based on many considerations. The expectation of making errors in determining relationships is built into the theory and practice of statistical analysis. The accuracy and potential of the statistics to support or predict a relationship among variables must be considered as a potential threat to the validity of a design.

Consider this example. In efforts to ensure that you do not overestimate the effects of your preventive program on AIDS risk behavior, you select a statistic that will be less likely to find a false relationship when there is none. However, this decision increases the possibility of overlooking a relationship when one exists. Thus, your statistical testing may lead you to believe that no relationship exists when there is one or, conversely, to claim a relationship when there is not one.

Construct Validity

Construct validity addresses the fit between the constructs that are the focus of the study and the way in which those constructs are operationalized. Because, as indicated earlier, constructs are abstract representations of what humans observe and experience, how one goes about defining and measuring constructs accurately is, in large part a matter of opinion and consensus. It is possible for several factors relating to construct validity to confound a study. First, it is possible for a researcher to define a construct inappropriately. For example, although smiling may be one indication of happiness, if a researcher measured level of happiness exclusively by observing the frequency of smiling behavior, the full construct of happiness would not be captured. Poor or incomplete operational definitions can result from incomplete conceptual definitions or from inadequate translation of the construct into an observable one. Second, when a cause and effect relationship between two constructs is determined, it may be difficult to exclusively define each construct (referring to both the independent and dependent variables) and to isolate the effects of one on the other.

Consider this example. Suppose you are an occupational therapist who is attempting to test a feeding program for persons with Alzheimer's disease. You design the program in which you choose certain foods and techniques to improve independent feeding. To ensure that your design can reveal cause and effect, you randomly select your sample and randomly assign them to the experimental group (which receives the program) and the control group (which does not receive the program). You then measure feeding behavior by two methods: observation of independent feeding and volume of food consumed independently. You implement your feeding program with the experimental group and then test all participants after the program to determine if the experimental group fared better on your measures of feeding than the group that did not receive the intervention. You conclude that your program was successful in improving independent feeding, because the experimental group performed so much better than the group without the intervention. However, you then begin to think about other confounding influences that cannot be separated from your two constructs; independent feeding and the experimental program. It is possible that the attention of the experimenter may have been responsible for the improvement in the experimental group, because eating is a social behavior and changes under different social conditions. Second, it is possible that food preferences influenced the response of the groups to their respective conditions. Interactive effects of being observed and participating in an experiment (referred to as the attention factor, or Hawthorne effect[7]) may also limit your capacity to isolate a causal relationship between the constructs.

Reliability

Reliability refers to the stability of a research design. In experimental-type research, rigorous research is so well planned and executed that if repeated under the same circumstances, the design should yield the same results. Investigators frequently "replicate," or repeat, a study in the same population or in different groups to determine the extent to which the findings of one study are "true" in a broader scope. For the study to be replicated, the procedures, measures, and data analytic techniques must be consistent, well articulated, and appropriate to the research question. Reliability is threatened when a researcher is not consistent and changes a design in midstream or does not fully plan a sound design to begin with.

SUMMARY

Design is a pivotal concept in the research process for experimental-type research in that it indicates both the structure and plan of action. Designs are developed to eliminate bias and the intrusion of unwanted factors that could potentially confound findings and make them less credible. We have defined basic terms used in structuring a design: concepts, constructs, conceptual and operational definitions, independent, intervening and dependent variables, and hypotheses. We then identified five primary considerations in the plan of a design, that of bias, control, manipulation, validity, and reliability. Finally, we demonstrated how four types of validity threaten study outcomes: internal, external, statistical conclusion, and construct.

Experimental-type research is designed to minimize the threats posed by extraneous factors and bias through maximizing control over the research action process. In the chapter that follows we discuss how each type of design on the experimental-type continuum addresses these considerations in structuring and planning a design and the variety of action processes that investigators use to increase control.

EXERCISES

1. Select a research article that uses an experimental-type design from a journal. First, identify the independent variable or variables, the dependent variable or variables and any intervening variables that might be confounding the study.

2. Identify the conceptual definitions and the operational definitions in the article that you selected.

3. Using the same article, determine (1) threats to validity of the study, (2) how the investigator minimized bias and enhanced control and validity, and (3) ethical issues that shaped the design.

REFERENCES

1. Kerlinger FN: *Foundations of behavioral research,* New York, 1973, Holt, Rinehart & Winston.
2. Wilson J: *Thinking with concepts,* Cambridge, Mass, 1966, Cambridge University Press.
3. Babbie E: *The practice of social research,* ed 6, Belmont, Calif, 1991, Wadsworth, p 118.
4. Polansky NA: *Social work research: methods for the helping professions,* Chicago, 1975, University of Chicago Press.
5. Goodman C: Evaluation of a model self help telephone program: impact of natural networks, *Soc Work* 35:556–562, 1990.
6. Campbell DT, Stanley JC: *Experimental and quasi-experimental design,* Chicago, 1963, Rand McNally.
7. Perrow C: *Complex organizations: a critical essay,* ed 2, New York, 1979, Random House.

8

Continuum of Experimental-Type Designs

- **True Experimental Design**
- **Variations of Experimental Designs**
- **Quasi-experimental Designs**
- **Pre-experimental Designs**
- **Non-experimental Designs**
- **Techniques for Increasing Control**
- **Criteria for Selection of Appropriate and Adequate Design**
- **Summary**
- **Exercises**

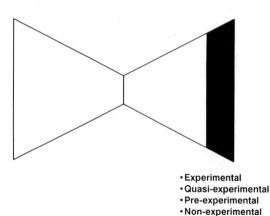

- Experimental
- Quasi-experimental
- Pre-experimental
- Non-experimental

Using the language introduced in Chapter 7, we are ready to examine the characteristics of designs included along the continuum of experimental-type research. In Chapter 6 we pointed out that designs along this continuum have traditionally been classified as either "true experimental," "quasi-experimental" or "non-experimental." As the language of this continuum implies, it is the *experiment* that has been the criterion by which all other methodologic approaches are judged. Of all the experimental-type designs, the "true experimental design" offers the greatest degree of control and internal validity, and therefore it is used to reveal causal relationships between independent and dependent variables.

Before we begin our discussion of experimental-type designs, it is important to recognize that each and every design along the experimental-type continuum has merit and value. The merit and value of a design is based on how well that design can answer the particular research question posed and on the investigator's rigor in the conduct and planning of the inquiry. This view of research differs from many experimental-type researchers who present the true experiment as the design of choice[1, 2] and suggest that other designs along this continuum are deficient or limited. However, we believe that a design along the experimental-type continuum should be chosen because it fits the question, level of theory development, and setting or environment in which the research will be conducted. In other words, not all research questions seek to predict causal relationships between independent and dependent variables. True experimentation is the best design to use to predict causal relationships, but it may be inappropriate for other forms of inquiry and research settings in health and human services.

Let us begin our discussion of designs on the experimental-type continuum with the true experimental design. To express the structural relationships of designs along this continuum, we use Campbell and Stanley's[3] notation system, which has been widely adopted by researchers. Stanley and Campbell used the following symbols to diagram design: X for independent variable, O for dependent variable, and R for random sample selection.

We also find it very helpful to use the symbol r to refer to random group assignment. As you will see in this and subsequent chapters, it is often very difficult and frequently inappropriate or unethical for health and human service professionals to select a sample from a larger predefined population based on random selection (referred to as R in Campbell and Stanley's[3] format). Rather, typically in health and human services research, subjects enter studies on a volunteer basis. Such a sample is one of convenience in which subjects are then randomly assigned to either the experimental or control group. The addition of r denotes this important structural distinction.

TRUE EXPERIMENTAL DESIGN

$$R \quad O \quad X \quad O$$
$$R \quad O \quad \quad O$$

The true experimental design is perhaps the design best known by beginning researchers and lay persons. True experimental design refers to the classic two group design in which subjects are randomly selected and randomly assigned (*R*) to either an experimental or control group condition. Before the experimental condition, all subjects are pretested or observed on a dependent measure (*O*). In the experimental group, the independent variable or experimental condition is imposed (*X*), and it is withheld in the control group. Subjects are then post-tested or observed on the dependent variable (*O*) after the experimental condition.

In this design, the investigator expects to observe no difference between the experimental and control groups on the dependent measure at pretest. That is, because subjects are chosen randomly from a larger pool of potential subjects and then assigned to a group on a chance-determined basis, subjects in both groups are expected to perform similarly. On the other hand, the investigator anticipates or hypothesizes that differences will occur between experimental and control group subjects on the post-test scores. However, this expectation is expressed as a null hypothesis, which states that "no difference" is expected. In a true experimental design, the investigator always states a null hypothesis that forms the basis for statistical testing. In research reports, however, usually the alternative hypothesis is stated (in other words, that of an expected difference). If the investigator's data analytic procedures reveal a "significant" difference between experimental and control group scores, he or she can fail to accept the null hypothesis with a reasonable degree of certainty. In failing to accept the null hypothesis, the investigator accepts with a certain level of confidence that the independent variable or experimental condition (*X*) *caused* the outcome in the experimental group observed at post-test time. In other words, the investigator infers that the difference at post-test time is not the result of chance alone but is due to participation in the experimental group.

There are three major characteristics of the "true experiment":

- Randomization
- Control
- Manipulation of an independent variable

Let us consider each of these characteristics.

Randomization

Randomization occurs at the sample selection phase and/or at the group assignment phase. If random sample selection is accomplished, the design notation appears as it was presented earlier (*R*). If randomization occurs only at the group assignment phase, we represent the design as such:

$$r \quad O \quad X \quad O$$
$$r \quad O \quad \quad O$$

This variation has implications for external validity. A true experimental design that does not use random sample selection is limited in the extent to which conclusions can be generalized to a larger population. Because subjects are not drawn by chance from a larger identified pool, the generalizability or external validity of findings is limited. However, such a design variation is very common in experimental research and can still be used to reveal causal relationships within the sample itself.

The random assignment of subjects to group conditions based on chance is essential in true experimentation. It enhances the probability that subjects in experimental and control groups will be equivalent on all major dependent variables at the pretest occasion. Randomization, in principle, equalizes subjects or makes subjects in both experimental and control groups comparable at pretest time. An observed change in the experimental group at post-test time can then be attributed with a reasonable degree of certainty to the experimental condition.

Randomization is a powerful technique that increases control and eliminates bias by neutralizing the effects of extraneous influences on the outcome of a study. For example, the threats to internal validity by historical events and maturation are theoretically eliminated. That is, based on probability theory, such influences should affect subjects equally in both the experimental and control groups. Without randomization of subjects, you would not have a true experimental design.

Control

We spoke about the concept of control in Chapter 7. Here we extend our discussion to refer to the inclusion of a control group. The control group allows the investigator to see what the sample would be like without the influence of the experimental condition or independent variable. A control group theoretically performs or remains the same at pretest and post-test occasions. It represents the characteristics of the experimental group before being changed by participation in the experimental condition.

The control group is also a mechanism that allows the investigator to examine what has been referred to as either the "attention factor," "Hawthorne effect," or "halo effect."[4] These three terms refer to the phenomenon that subjects may experience change just from the act of participating in a research project. For example, just being the recipient of personal attention from an interviewer during pretesting and post-testing times may influence how subjects feel and respond to interview questions. A change in scores may occur then, independent of the effect of the experimental condition. Without a control group, investigators would not be able to say that differences on post-test scores reflect that of experimental effect and not just additional attention.

Interestingly, this attention phenomenon was discovered in the process of conducting research. In 1934, a group of investigators were examining productivity in the Hawthorne auto plant in Chicago. The research involved interviewing workers. To improve productivity, the investigators recommended that the lighting of the facility be brightened. The researchers noted week after week that productivity increased after each subsequent increase in illumination. To confirm the success of their amazing findings, the researchers then dimmed the light. To their surprise, productivity continued to increase even under this circumstance. In re-examining the research process, they concluded that it was the additional attention given to the workers through the ongoing interview process and their inclusion in the research itself that had caused an increased work effort.[5]

Manipulation

In the true experimental design, the independent variable is manipulated by either having it present (in the experimental group) or absent (in the control group). It is the ability to provide and withhold the independent variable that is unique to the true experiment.

According to Campbell and Stanley,[3] theoretically, true experimentation controls for each of the major seven threats to internal validity discussed in Chapter 7. However, when such a design in the health care or human service environment is actually implemented, certain influences may remain as internal threats, thereby decreasing the potential for the investigator to support a cause-effect relationship. For example, the selection of a data collection instrument in which there is a learning effect based on repeated testing could pose a significant threat to a study regardless of how well the experiment is structured. It is also possible for experimental mortality to affect outcome, particularly if the groups become nonequivalent as a result of attrition from the experiment. The health or human service environment is complex and does not offer the same degree of control as a laboratory setting. Therefore, in applying the true experimental design to the health or human service environment, the researcher must carefully examine how each threat to internal validity is or can be resolved.

VARIATIONS OF EXPERIMENTAL DESIGNS

There are many design variations of the true experiment that have been developed to enhance its internal validity. For example, to assess the effect of the attention factor, a researcher could develop a three-group design. In this design, subjects are randomly assigned to either an experimental condition, an attention control group that receives an activity designed to equalize the attention subjects receive in the experimental group, or a silent control group that receives no attention other than that obtained naturally during data collection efforts. Silent control groups can also refer to the situation in which information is collected on subjects who have no knowledge of their own participation. For example, extracting information from medical records or other such artifacts on a group who remain unaware may provide the researcher with an understanding of how subjects in their study compare with those who are not, at least along key demographic and medical characteristics.

Let us examine four other basic design variations of the true experiment.

Posttest-Only Designs

Posttest-only designs conform to the norms of experimentation in that they contain the three elements of random assignment, control, and manipulation. The difference between classical true experimental design and this variation is the absence of a pretest. The design notation for a posttest-only experiment is:

$$R \quad X \quad O$$
$$R \qquad\quad O$$

In this design, a leap of faith, so to speak, is taken in that groups are considered to be equivalent before the experimental condition as a result of random assignment. Theoretically, randomization should yield equivalent groups. However, the absence of the pretest makes it impossible to determine if random assignment successfully achieved equivalence between the experimental and control groups on the major dependent variables of the study. Posttest-only designs are most valuable when pretesting is not possible or appropriate but the research purpose is to seek causal relationships.

Solomon Four-Group Design

The Solomon four-group design represents a more complex experimental structure. It combines the true experiment and the post-test-only design into one design structure. The strength of this design is that it eliminates the potential influence of the test-retest learning phenomenon by adding the post-test-only two-group design to the true experimental design. This design is noted as:

$$
\begin{array}{llll}
\text{Group 1:} & R & O & X & O \\
\text{Group 2:} & R & O & & O \\
\text{Group 3:} & R & & X & O \\
\text{Group 4:} & R & & & O
\end{array}
$$

As you can see by this notation, group 3 does not receive the pretest but participates in the experimental condition. Group 4 also does not receive the pretest but serves as a control group. By comparing the posttest scores of all groups, the investigator can evaluate the effect of testing on scores on the post-test and the interaction between the test and the experimental condition. The key benefit of this design is its ability to detect *interaction effects*. An interaction effect refers to changes that occur in the dependent variable as a consequence of the combined influence or interaction of taking the pretest and participating in the experimental condition.

Let us consider an example to illustrate the power of the Solomon four-group design and the nature of interaction effects. Suppose you wanted to assess the effects of an AIDS information training program (independent variable) on the sexual risk-taking behaviors of adolescents (the dependent variable). You would pretest groups by asking them questions regarding their sexual activity and level of knowledge regarding risks of obtaining AIDS. Then you would expose one group to the experimental condition that involves attending a peer-led educational forum. On post-testing you find that levels of knowledge increased and that risk behaviors decreased in subjects who received the experimental program. However, you would not be able to determine the effect of the pretest itself on the outcome. By adding group 3 (the experimental condition without a pretest), you can determine if the change in scores is as strong as when the pretest is administered (group 1). If group 2 (control) and group 3 (experimental) show no change but experimental group 1 does, this change is a consequence of an interaction effect of the pretest and the intervention. If there is a change in experimental groups 1 and 3 and some change in control group 2 but none in control group 4, there may be a direct effect of the intervention or experimental condition plus an interaction effect. This design allows the investigator to determine the strength of the independent effects of the intervention and the strength of the effect of pretesting on outcomes. If the groups that have been pretested show a testing effect, statistical procedures can be used to correct for it if necessary.

As you can see, the Solomon four-group design offers increased control and the possibility of understanding complex outcomes. However, because it requires the addition of two groups, it is a costly and time-consuming alternative to the true experimental design and is not frequently used in health and human service inquiry.

Factorial Designs

Factorial designs offer even more opportunity for multiple comparisons and complexity in analysis. In these designs the investigator evaluates the effects of either two or more independent variables (X_1 and X_2) or the effects of an intervention on different factors or levels of a sample or study variables. The interaction of these parts, as well as the direct relationship of one part to another, is then examined.

Let us say you were interested in examining the effects of an exercise program for older adults. You want to determine the extent to which two levels of health status (good and poor) influenced three areas of quality of life for participants. In this case, the first independent variable that is manipulated assumes two values: participation in exercise or participation in a nonexercise control group. The other independent variable that is not manipulated (health)

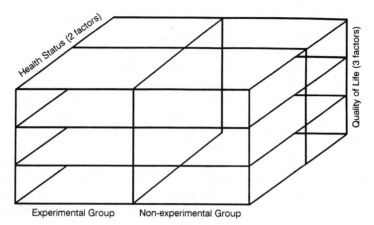

FIG 8–1.
Factorial design: investigating state of quality of life in elderly participants in experimental exercise program.

also has two values (good and poor). The dependent variable or outcome measure (quality of life) has three factors (let us say, in this example, these factors are activity level, overall satisfaction, and sense of well-being). This study represents a factorial design in which the independent variables have two factors or levels and the dependent variable has three levels. This structure is referred to as a 2 X 2 X 3 factorial design, as displayed in Figure 8–1. This design would allow you to examine the relationship between different levels of health and specific quality of life indicators for two different conditions: one in which subjects exercised and one in which they participated in a nonexercise control group. This design would enable you to examine not only direct relationships but interactive relationships as well. You could determine the combined effects of two or more variables that may be quite different from each direct effect. For example, you might want to explore if the less healthy older exerciser benefited more on the activity indicator of quality of life than the less healthy older nonexerciser and so forth. To statistically evaluate all of the possible contrasts in this design, you would need to have a very large sample size to assure that each cell or block in Figure 8–1 was adequate. Of course, you can develop much more complex factorial designs in which both independent and dependent variables have more than two levels.

Counterbalance Designs

Counterbalance designs are often used when more than one intervention is being tested and when the order of participation is manipulated. This design allows the investigator to determine the combined effects of two or more interventions and the effect of order on study outcomes. Although there are many variations, the basic design is:

$$R \quad O \quad X_1 \quad O \quad X_2 \quad O$$
$$R \quad O \quad X_2 \quad O \quad X_1 \quad O$$

Note the subscript numbers on the independent variables. The reversal of the conditions is characteristic of counterbalance designs in which both groups experience both conditions but in different orders. This reversal is called crossover. In a crossover you assign one group to the experimental group first and to the control condition later and then reverse the order for the other group. Measurements would occur before the first set of conditions, before the second set, and after the experiment. Such a study would allow you to eliminate the threats to internal validity caused by the interaction of the experimental variable and other aspects of the study.

Goodman's study of the psychosocial effects of two different interventions for caregivers of individuals with Alzheimer's disease is an example of a counterbalance study.[6] Goodman randomly assigned her sample to one of two conditions, lecture only and peer telephone network, and then reversed their assignment. Testing occurred at three intervals: just before the experiment, after participation in condition one, and after the experiment. Goodman was not only able to ascertain the effects of each program on a single group but was also able to compare groups over time in each of the conditions.

Summary

True experimental designs must contain three essential characteristics: random assignment, control, and manipulation. The classic "true experiment" and its variations all contain those elements and thus are capable of producing knowledge about causal relationships among independent and dependent variables. The four design strategies in the "true experimental" classification are appropriate for an experimental-type research question in which the intent is to predict and reveal a "cause." Each of these true experimental types controls the influences of the basic threats to internal validity and unwanted or extraneous phenomena that can confound a causal study and invalidate causal claims. There are three criteria for using a true experimental design or its variation: sufficient theory to examine causality; a causal question; and conditions and ethics that permit randomization, control group, and manipulation.

QUASI-EXPERIMENTAL DESIGNS

As we have indicated throughout the book, although "true experiments" have been upheld as the ideal or prototype in research, there are many reasons why such designs may not be appropriate. First, in health and human service research, it may not be possible, appropriate, or ethical to use randomization or to manipulate the introduction and withholding of an experimental intervention. Second, all inquiries do not ask causal questions. There are other design options that do not contain the three elements of a true experiment that are frequently used by health and human service researchers to generate valuable knowledge. Even though the language on the experimental-type continuum implies that these designs are "missing something," we suggest that these designs are not inferior but, rather, produce different interpretations and uses than true experimentation. Decisions to use quasi-experimental design should be based on the level of theory, type of research question asked, and constraints of the research environment.

According to Cook and Campbell,[7] who wrote the seminal work on quasi-experimentation, these designs can be defined as:

> experiments that have treatments, outcome measures, and experimental units, but do not use random assignment to create comparison from which treatment caused change is inferred. Instead, the comparisons depend on non-equivalent groups that differ from each other in many ways other than the presence of the treatment whose effects are being tested [p. 6].

The key to the effective use of quasi-experimental designs lies in the claims that the researcher makes about the findings. Because random assignment is absent in quasi-experimentation, the researcher has one of two choices: (1) to make causal claims while acknowledging the alternative explanations for those claims and design limitations or (2) to avoid making causal inferences when they are unjustified by the design.

As discussed by Campbell and Cook,[7] designs belonging to this category have two of the three experimental elements: control and manipulation. Here we discuss three basic design types that fit the criteria for quasi-experimentation:

- Nonequivalent control group
- Interrupted time series
- Single subject designs

Nonequivalent Control Group

In the nonequivalent comparison group design, there are at least two comparison groups, but subjects are not randomly assigned to these groups. The basic design is structured as a pretest, post-test comparison group, as presented in this notation:

$$O \quad X \quad O$$
$$O \qquad O$$

It is also possible to add comparison groups or to alter the testing sequence. Let us consider an example in which you want to test the effectiveness of an innovative mental health program for depression but it is not possible to use randomization. You arrange for a community mental health center to use the experimental intervention for 1 month. You find another community mental health center in which the population is comparable and assign the comparison condition (conventional intervention) to that group. As with true experimental design, you pretest and post-test all subjects and then compare group scores. If the group scores for the experimental condition are significantly different, there is strong support for the value of the innovative program. However, once again, because of the multiple threats to internal validity, you would most likely not use this design to support cause but to explain the comparative changes in each of the study groups. That is, you can indicate that the changes after the intervention in the experimental group were significantly greater than the changes in the comparison group.

Interrupted Time Series

Interrupted time series designs involve repeated measurement of the dependent variable both before and after the introduction of the independent variable. There is no control or comparison group in this design. The multiple measures before the independent variable control for the threat to internal validity based on maturation and other time related changes. A typical time series design is depicted as:

$$O_1 \quad O_2 \quad O_3 \quad X \quad O_4 \quad O_5 \quad O_6$$

Although the number of observations may vary, it is suggested that no less than three occur before and after the independent variable is introduced. The investigator is particularly interested in evaluating the change in scores between the observation that occurs immediately before the introduction of the intervention (O_3) and the observation that follows the intervention (that of O_4). Any sharp change in score compared with the other measures may suggest that the intervention had an effect. The investigator must evaluate changes in scores in terms of the scoring patterns that occurred both before and after intervention. For example, if there is a trend for scores to increase at each testing occasion, even a sharp difference between O_3 and O_4 may not reflect a change because of the introduction of the intervention. Because of the absence of the control group and randomization, this design cannot strongly support a causal relationship between the independent and dependent variables. In this type of design, however, the series of premeasures theoretically control threats to internal validity except for that of history. Maturation, testing, instrumentation, regression, and attrition are considered threats that may occur between all measures and thus are detectable and controlled for by repeated measurements. However, in this design, it is extremely important to choose a form of measurement in which there is no learning effect or threat of testing.

In the health and human service context, a time series design is often used when the experimental intervention is a stimulus the researcher may not be able to manipulate but has knowledge of its introduction. That is, the intervention or experimental program may be a naturally occurring event in which it is possible to document performance of the dependent variable before and after the event's occurrence. For example, let us say a hospital announces a plan to implement a new employee benefits program to enhance job satisfaction. The effects of this program on job satisfaction could be determined by taking a quarterly survey of employees for 1 year before the introduction of the new program. Then the same survey could be used quarterly after employee participation in the new benefits program. In this way, the hospital would have four data points regarding employee satisfaction before the program's introduction that could be compared with four data points after its implementation. This would allow for such extraneous factors as staff turnover, fluctuations in patient census, and other considerations to be tracked over time to account for their effect on job satisfaction. This type of design is extremely useful in answering questions about the nature of change over time.

A combination of nonequivalent groups and time series design is a quasi-experimental design that can be considered to be appropriate for answering causal questions. In this combination, some of the potential bias attributable to nonequivalent groups is limited by multiple measures, whereas the time series design is strengthened by the addition of a comparison group.

Single-Subject Design

The time series design structure is also frequently used by researchers when they are following a single subject over time. Suppose, for example, you wanted to determine if an innovative intervention with a chronically depressed population decreased levels of depression. Because you are unable to test a group of subjects, you select a single-subject design. For the first three sessions, you administer a depression inventory. You then conduct the intervention and follow up with three administrations of the same instrument. If, based on visual inspection, you observe a difference between the observed scores immediately preceding and immediately after the intervention, you can surmise that a change occurred and that this change may be a consequence of the intervention. With this type of design you cannot claim that the intervention was the cause of the improvement. However, you have demonstrated a change, have a stronger case for the possibility of a causal relationship, and have support for continuing to investigate the intervention with another design.

There are many variations of designs that are classified as single-subject, single-system, or N of 1 single case experimental trial.[8-11] These designs involve a single individual or series of single individuals and are useful in the following circumstances: (1) when it is not possible to randomize, (2) when it is not possible to study a particular clinical population as a group with similar characteristics, (3) to determine clinical outcomes or change in a behavior over time as a consequence of therapeutic treatment or multiple interventions, or (4) to obtain pilot information in a cost-efficient way.

Single-subject research has become an increasingly popular strategy to systematically assess the outcomes of therapeutic approaches for a single client. There are many different strategies in single-subject research, but each is based on the premise that the individual serves as his or her own control. The most basic or elementary design is called *AB* design, where *A* represents baseline measurements and *B* the intervention that is introduced. During the baseline period (*A*), repeated measures are obtained on specified variables. It is suggested that the investigator obtain as many measurements as possible (six to nine observations, for example) to establish a stable baseline. However, the number of measurements may vary based on whether the observed behavior is stable. If, for example, observed scores are variable, many repeated measures may be necessary to detect the baseline pattern of behavior before the intervention phase. These measures form the "control" or base from which measures after the intervention phase (*B*) are compared. It is also recommended that a similar number of observations be obtained in the *B* phase to examine the stability of change over time. After the *AB* assessments, the investigator plots each score and then performs a visual analysis to detect a change in the pattern of scoring between the *A* and *B* phases. Typically a "celeration" line is drawn through the baseline data to facilitate visual inspection of observed scores. This line represents the best fit of the baseline data points. The same line is then drawn along the data points in phase *B* to examine if there is an apparent difference in the direction and placement of scores.

Variations on the *AB* design include the *ABA* and *ABAB* designs. In the *ABA*, there is a baseline period (*A*), followed by an intervention period (*B*), which is then followed by the withdrawal of the intervention (*A*). In the *ABAB* design, the previous sequence is followed and then the intervention is reapplied. There are many variations of even these design structures. For example, each *B* phase may involve different therapeutic strategies.

A major concern in this type of research is that of generalizability. Researchers involved in single-subject research suggest that limited external validity is overcome through generalization to theory, replication, or the repetition of systematic single-subject studies, and the evaluation of each of these efforts as a whole.[12]

Summary

Quasi-experimental designs are characterized by the presence of control and manipulation, but do not contain random group assignment. In the time series and single-subject designs, although there is no control group, control is exercised through multiple observations of the same phenomenon both before and after the introduction of the experimental condition. In non-equivalent group designs, the control is built in through the use of one or more comparison groups. We suggest that quasi-experimentation is most valuable when the investigator is attempting to look for change over time or a comparison between groups and the realities of the health or human service environment are such that random assignment is not appropriate, ethical, or feasible.

PRE-EXPERIMENTAL DESIGNS

Pre-experimental designs are those in which two of the three criteria for true experimentation are absent. In pre-experiments, it is possible to describe phenomena or relationships. However, the outcomes of the study do not support claims for a causal relationship because of inadequate control and the potential of bias. Pre-experimental designs can be of value to answer descriptive questions or to generate pilot, exploratory evidence, but the investigator using these designs cannot consider causal explanations. There are numerous preexperi-mental designs, all of which are variations on the following three designs:

- One-shot case study
- Pretest/post-test designs
- Static group comparison

One-Shot Case Study

The one-shot case study is diagrammed as:

$$X \quad O$$

In this case, the independent variable is introduced and the dependent variable is then measured in only one group. Without a pretest or a comparison group, the investigator can answer the question "How did the group score on the dependent variable after the intervention?" As you can see, a cause-and-effect relationship between the two variables cannot be supported because of the seven threats to internal validity.

Pretest-Posttest Design

Similarly, the pretest-posttest design, depicted below, is also valuable in describing what occurred after the introduction of the independent variable.

$$O \quad X \quad O$$

This design can answer questions about change over time in that the pretest is given before the introduction of the independent variable. If subjects are tested before the intervention and after the intervention, a change in scores on the dependent variable can be reported but cannot be attributed to the influence of the independent variable. Threats to internal validity, if one were to attempt to infer cause using this design, include maturation, history, testing, instrumentation, experimental mortality, and interactive effects.

Static Group Comparison

In the static group comparison, a comparison group is added to the one-shot case study design:

$$X \quad O$$
$$O$$

This design, as in the other pre-experimental designs, can answer descriptive questions about phenomena or relationships, but is not what we consider to be a desirable choice for causal studies. Let us say we wanted to test an intervention that is designed to reduce depression in young adults. The investigator who uses the static group comparison would select two nonequivalent groups, such as two groups of persons with depression receiving treatment in two different community mental health centers. One group would receive the intervention and one would not. A post-test measuring level of depression would be administered to both groups and then compared. This design could answer the following question with a fair degree of certainty: "How did the experimental group compare with the comparison group on the measure of depression?" Because random assignment did not occur, it is difficult to infer cause. Furthermore, there is minimal control in that the level of depression was unknown before the introduction of the experimental condition in either group. However, by introducing a static comparison group, this design offers more control over extraneous factors than the one-shot case and pretest/post-test designs.

Summary

Pre-experimental designs may be valuable in answering a descriptive question. However, the absence of two of the three major conditions for true experimentation makes these designs an inappropriate choice if one's pursuit is prediction and causal inference. If one attempts to answer predictive or causal questions with these designs, to the basic seven threats to internal validity limit effective checks against bias.

NON-EXPERIMENTAL DESIGNS

Non-experimental designs are those in which the three criteria for true experimentation do not exist. These designs can range in use from exploration to prediction and are most useful when testing a concept or construct or relationship among constructs. For the most part, non-experimental designs examine naturally occurring phenomena and describe or examine relationships. Any manipulation of variables is done post hoc through statistical analysis. We describe three designs that are frequently used in health and human service research:

- Surveys
- Passive observation
- Ex post facto designs

Survey

Survey designs are used primarily to measure characteristics of a population. Through survey design it is possible not only to describe population parameters but to predict relationships among those characteristics as well. Typically surveys are conducted with large samples. Questions are posed through either mailed questionnaires or telephone or face-to-face interviews. Perhaps the most well-known survey is the U.S. Census, in which the government administers mailed surveys and conducts selected face-to-face interviews to develop a descriptive picture of the characteristics of the population of the United States.

The study of job satisfaction of occupational therapy faculty conducted by Rozier et al.[13] is an example of how a survey design was used in health care not only to describe a population but also to predict factors that could lead to job satisfaction. The investigators mailed questionnaires to program directors of schools of occupational therapy and asked the directors to distribute these instruments to faculty. Five hundred thirty-eight faculty responded, yielding information about demographic characteristics and job satisfaction. Through statistical analysis of the data, descriptive and predictive conclusions were developed.

The advantages of survey design are that the investigator can reach a large number of respondents with relatively minimal expenditure, numerous variables can be measured by a single instrument, and statistical manipulation during the data analytic phase can permit multiple uses of the data set. Disadvantages may include how the action processes of the survey design are structured. For example, the use of mailed questionnaires may yield a low response rate, compromising the external validity of the design. Face-to-face interviews are time consuming and may pose reliability problems.

Passive Observation

Passive observation designs are used to examine phenomena as they naturally occur and to discern the relationship between two or more variables. Often referred to as "correlational designs," passive observation can be as simple as examining the relationship between two variables, for example, height and weight, or can be as complex as predicting scores on one or more variables from knowledge of scores on other variables. As in the case of the survey, variables are not manipulated but are measured and then examined for relationships and patterns of prediction. For example, in the survey just described by Rozier et al.,[13] the investigators examined the relationships among multiple variables and predicted the degree of teaching satisfaction with respondent scores on four other variables.

Ex Post Facto Designs

Ex post facto designs are considered to be one type of passive observation design. However, in ex post facto design (literally translated as after the fact), the phenomena of interest have already occurred and cannot be manipulated in any way. Ex post facto designs are frequently used to examine relationships between naturally occurring population parameters and specific variables. For example, let us say you were interested in understanding the effects of coronary by-pass surgery on morale and resumption of former roles for males and females. Coronary by-pass surgery is the event that the researcher cannot manipulate but that can be examined for its effects after its occurrence. Fortune and Hanks[14] used an ex post facto survey design to examine the differing career patterns of male and female social workers. Through surveying 520 recent graduates of schools of social work about their job history and their gender, Fortune and Hanks were able to answer questions about gender differences in salary and career opportunity in the field of social work.

Summary

Non-experimental designs have a wide range of uses. The value in these designs lies in their ability to examine and quantify naturally occurring phenomena, so that statistical analysis can be accomplished. In this category of design the investigator therefore does not manipulate the independent variable but examines it in relation to one or more variables for descriptive or predictive purposes. These designs have the capacity to include a large number of subjects and to examine events or phenomena that have already occurred. Because random selection, manipulation, and control are not present in these designs, investigators must use caution when making causal claims from the findings. As in the quasi-experimental and true experimental situation, the researcher is still concerned with potential biases that may limit the internal validity of the design. The researcher tries to control the influence of external or extraneous influences on the study variables through the implementation of systematic data collection procedures, the use of reliable and valid instrumentation, and other techniques discussed later. The researcher also tries to increase the generalizability of a study or its external validity by using random sample selection procedures when appropriate and feasible to assure representation and minimize systematic sampling bias.

TECHNIQUES FOR INCREASING CONTROL

Researchers use a variety of techniques to increase control and eliminate bias in the designs along the experimental-type continuum. We have listed seven basic techniques:

- Randomization
- Matching
- Blocking
- Analysis of covariance
- Sample size
- Homogeneity in selection of sample
- Standardization of procedures

Some of these techniques we have already discussed, for example, randomization. However, let us review each and describe how it increases control, eliminates bias, and increases the validity of studies.

Randomization

As we discussed previously, "random" in the experimental-type world does not mean haphazard. Rather, it refers to probability theory and is defined as sample inclusion and group assignment on a chance-determined basis. Thus each member of the population has an equal chance of being selected for a sample or each member of the sample has an equal chance of being selected for either the experimental or control group condition. If a sample is randomly selected and assigned, the potential for bias is reduced significantly, because there are no "unwanted" reasons for selection or assignment. Furthermore, theoretically each unit of analysis has an equal chance of being exposed to the same extraneous influences. This equal opportunity decreases the potential for extraneous influences to effect one group and not another or to effect one group in a different way than the other. Usually a table of random numbers is used to assign subjects to groups.

Matching

If, however, random selection of a sample is not possible or random selection and assignment may not accomplish equivalent groupings, matching can be an excellent technique to reduce potential bias. Matching is often used in quasi-experimental designs to reduce the potential influence of unwanted subject characteristics on the outcomes of a study. In matching, the researcher must determine which variables or characteristics of a sample are related to the dependent variable. The researcher selects subjects with specific characteristics and then assigns matched pairs to different study groups. This matching is an attempt to achieve equivalence in subject characteristics between two or more groups. Let us say you are doing a study examining the effect of an exercise weight-lifting program on strength building. You have determined through a review of the literature that age and sex are two variables that will influence performance on your dependent variable, strength. Although you cannot randomly assign subjects, you match subjects by their gender and age and then assign one from each pair to the experimental condition and the other to the control group. In this way you have made your two groups similar so that you can compare their performance before and after the experimental condition. If you had not matched, you might have had unequal groups. For example, suppose the experimental group was composed of younger males and the control group had either older males or females. The research literature indicates that younger males demonstrate greater strength capacity than either of the subject types in the control group. The differences you might have found would therefore be confounded and biased.

Blocking

Although random assignment theoretically produces equivalent groups, this may not always be the case. There are occasions when randomization, just by chance, may result in nonequivalent groups. Nonequivalence can be a significant problem with small sample sizes. To avoid the chance that randomization will not produce equal groups, some researchers use a technique called blocking, which is a combination of randomization and matching. As in the case of matching, the researcher blocks on variables that may have a potential effect on the dependent variable. To block, subjects are first matched on one or more salient variables. Then members of each block are randomly assigned to the experimental and control group conditions. Blocking provides additional assurance that extreme scores will be evenly distributed among the groups and that the groups will be equivalent on the key blocking variables.

Analysis of Covariance

Analysis of covariance is a statistical technique that removes the effect of the influence of another variable on the dependent or outcome variable. This statistical test mathematically corrects for any influences on the dependent variable that are a function of the respondent's pretest scores on post-test scores or any other identified factor in which it is important to control its effect.

Sample Size

Sample size is a complex concept. Although many beginning researchers believe that the larger the sample the less chance of bias and error, this may not always be the case. Sample size is dependent on the research question, the field constraints, and the selection of data analytic techniques. We discuss sample size more fully in Chapter 8. Suffice it to say in this introduction that sample size should be carefully considered and planned so that bias is minimized while external validity and control are maximized.

Homogeneity

As we indicated earlier, as the sample becomes more complex in its characteristics, the potential for confounding factors resulting from sample characteristics increases. Although techniques such as statistical tests that correct for covariance and matching the sample selection and assignment may be useful, bias can be minimized by narrowing the population. Selecting a homogeneous sample or subjects who are very similar along select character-istics is another technique that is sometimes used. For example, in a study to test a caregiver intervention, Toseland et al.[15] restricted their sample to adult daughters and daughters-in-law who were primary caregivers for their relatives. In this way, the effects of the intervention were examined for a sample that shared similar experiences, roles, and familial responsibilities. This restriction theo-retically increased the power of the study to detect differences for this specific population. If other caregivers such as males or spouses had been included, gender and role may have confounded or influenced study outcomes.

Standardized Procedures

Finally, standardizing procedures is an important mechanism for increasing control. Just think about what could happen if each interviewer in a survey asked questions in a different way. Because the design in experimental-type research is developed and finalized before entering the field or engaging in the action of data collection, the standardization of techniques can be built into a protocol in which interviewers and other members of the research team are thoroughly trained to enact.

CRITERIA FOR SELECTION OF APPROPRIATE AND ADEQUATE DESIGN

As discussed earlier in this chapter, there is a very strong belief in the research world that the true experiment represents the only design that is appropriate and adequate. However, we hope by now you can see that each design along the continuum of experimental-type design has its strengths and limitations. The true experiment is the best design for testing theory and making causal statements. However, if causality is not your purpose or if the design structure does not fit the particular environment in which the research is to be conducted, it is not an appropriate or adequate design choice.

It is often difficult to apply strict experimental conditions to a field setting. For example, although subjects may be randomly assigned to a group, it may not be possible to obtain the initial list of subjects for a random sampling process. Or it may be unethical to withhold a type of treatment (or experimental intervention) from a service consumer. Although clinical drug trials or testing of new technologies often obtain the degree of control necessary for true experimental conditions, research on the social and psychologic dimensions of health and human service work often pose a different set of issues and challenges for the researcher. These challenges make it essential for the investigator to be flexible in the use of design so that a research strategy appropriate to the question, level of theory, purpose, and practical constraints can be selected and rigorously applied.

To illustrate the value of each level of design on the experimental-type continuum, let us consider an example. Suppose you are employed in a hospital in the rehabilitation unit. You have just read about a new computer intervention that seems to enhance the cognitive recovery of persons with traumatic brain injury. You order the program and recruit a group of patients to use it. To determine the extent to which the program works (which you have defined as significant improvement in cognitive function), you select an instrument to measure cognitive function (operationalize your concept of cognitive improvement) and then test a sample of ten patients with that instrument after their use of the computer intervention. This type of inquiry is an *XO* design (pre-experimental) in which the computer program is the independent variable and cognitive performance is the dependent variable. The mean score for your sample was within normal limits, with a normal dispersion of scores around the mean. From this design, you have answered the question "How did subjects score on a test of cognitive performance after their participation in a computer intervention?" Although these data are valuable to you in describing your sample, you still have not answered the question about whether the program works. There are multiple threats to internal validity that interfere in your ability to make a causal inference between variables with any degree of assuredness. You therefore decide to build on this study by adding a pretest. In the next study you pretest your new sample, introduce the computer intervention, and then post-test the sample. This type of study is an *OXO* design (still pre-experimental). Because the subjects have greatly improved from the pretest to the post-test, you conclude that your program has worked. However, can you really do that? What about the effects of sampling bias, maturation, history, and so forth on your sample? You realize that even though you have stronger evidence for the value of the program, the design that you have selected has answered the question "To what extent did the subjects change in cognitive function after participating in the experimental condition?"

For your next sample, you add a control group so that your research design now is a quasi-experimental design and looks like this:

$$O \quad X \quad O$$
$$O \qquad O$$

When the experimental group improves much more than the control group, you now are convinced that your program works. However, can you make that claim? What about sampling bias and other interactive effects that might confound your study? From this design, you can answer the question "Which group made more progress?" This type of design provides fairly strong support for the value of your program but cannot yet be used to make causal claims.

For your final project, you add random group assignment to your design so that you are conducting a true experiment. Your design looks like this:

$$R \quad O \quad X \quad O$$
$$R \quad O \qquad O$$

Because your experimental group has improved significantly more than the control group, you can now make the claim, with reasonable certainty, that the program worked.

As you can see, each level of design has its values and limitations not only related to claims that can be made about the knowledge generated but about practical and ethical issues as well. The key to doing rigorous and valued research with experimental-type designs is to clearly state your question, select a design that can best answer the question given the level of theory development, the practical considerations, and the purpose in conducting the study, and report accurate conclusions.

SUMMARY

Now that you are aware of experimental-type design structures and techniques for enhancing the strength of your design, you still may wonder how to select a design to fit your particular study question. Begin by asking yourself the following important guiding questions:

1. Does the design answer the research question? Is there congruence between the research question or questions, hypotheses, and the design?
2. Does the design adequately control independent variables? That is, are other extraneous independent variables present that may confound the study?
3. Does the design maximize control and minimize bias?

4. To what extent does the design enhance the generalizability of results to other subjects, other groups, and other conditions?
5. What are some of the ethical and field limitations of the research question that influence the research design?

What do each of these questions really ask you to consider? First, in question 1, the issues of validity that were discussed in the previous chapter are raised. When a design is selected, it is essential that the project be internally consistent, construct valid, and capable of providing the level of answer that the question seeks. Question 2 also speaks to validity but addresses issues of control and structure as well. This question guides the researcher to consider what extraneous influences could potentially confound the study and what ways can maximum control be built into the design. Question 3 summarizes the issues raised in the previous questions by guiding the researcher to be most rigorous in planning control and minimizing bias. External validity and sampling are the focus of question 4. The answer to this question should be the development of a design to allow as broad a generalization as possible without threatening internal validity. Finally, in question 5, the researcher is reminded of practical and ethical concerns that shape the design.

If the questions are answered successfully, the criteria for adequacy of experimental-type designs have been addressed. However, keep in mind that design considerations along the experimental-type continuum are more than just knowing the definitions of design structures. Knowledge of these design possibilities is your set of tools, which you then apply and modify to fit the specific conditions of the setting in which you are conducting research. Developing a design along the continuum of experimental-type research in the health and human service field is a creative process. You must use your knowledge of the language and thinking processes, the basic elements of design, and the issues of internal and external validity to develop a study that fits the particular environment and question that you are asking.

EXERCISES

1. Select an experimental-type research article and identify the dependent and independent variables. Now diagram the design using *XO* notation.
2. Develop research questions and designs using the *XO* notation to illustrate:
   a. true experimental design
   b. quasi-experimental design
   c. pre-experimental design
   d. non-experimental design

3. Identify up to three potential ethical and field limitations of each research design you have developed in exercise 2.

REFERENCES

1. Kerlinger FN: *Foundations of behavioral research,* ed 2, New York, 1973, Holt, Rinehart & Winston.
2. Currier DP: *Elements of research in physical therapy,* Baltimore, 1984, Williams & Wilkins.
3. Campbell DT, Stanley JC: *Experiment and quasi-experimental designs for research,* Chicago, 1963, Rand McNally.
4. Perrow C: *Complex organizations: a critical essay,* ed 2, New York, 1979, Random House.
5. Roethlisberger FJ, Dickson WJ: *Management and the worker,* Cambridge, Mass, 1947, Harvard University Press.
6. Goodman C: Evaluation of a model of self help telephone program: impact of natural networks, *Soc Work* 25:556–562, 1990.
7. Cook TD, Campbell DT: *Quasi-experimentation: design and analysis for field settings,* Boston, 1979, Houghton Mifflin.
8. Giannini E: The N of 1 trials design in the rheumatic diseases, *Arthritis Care Res* 1:109–115, 1988.
9. Case-Smith J: An efficacy study of occupational therapy with high-risk neonates, *Am J Occup Ther* 42:499–506, 1988.
10. Hacker B: Single subject research strategies in occupational therapy, part 1 (research methods, case study), *Am J Occup Ther* 34:103–108, 1980.
11. Hacker B: Single subject research strategies in occupational therapy, part 2 (case study, research method), *Am J Occup Ther* 34:169–175, 1980.
12. Ottenbacker KJ, Bonder B: *Scientific inquiry design and analysis issues in occupational therapy,* Rockville, Md, 1986, American Occupational Therapy Foundation.
13. Rozier C, Gilkeson G, Hamilton BL: Job satisfaction of occupational therapy faculty, *Am J Occup Ther* 45:160–165, 1991.
14. Fortune AE, Hanks LL: Gender inequities in early social work careers, *Soc Work* 33:221–226, 1988.
15. Toseland RW, Rossiter CM, Peak T, et al: Comparative effectiveness of individual and group interventions to support family caregivers, *Soc Work* 25:309–317, 1990.

9

Naturalistic Inquiry: Language and Thinking Processes

- **Elements of Thinking Process**
- **Summary**
- **Exercises**

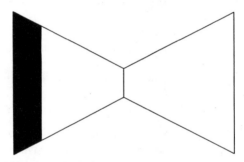

PLURALISTIC LANGUAGE AND THINKING PROCESSES

Now we would like you to set aside and suspend the vocabulary, procedures, and "ways of doing" research along the experimental-type continuum that we have just examined in the last two chapters. As you continue to read Chapters 9 and 10, we no longer want you to think of "bias," "control," "internal validity," or "manipulation." Come with us as we enter a new world of research, a different language, set of principles, and way of thinking.

Let us begin our understanding of naturalistic designs with a discussion and review of the basic orientations and thinking processes that underlie the continuum of naturalistic inquiry. As we have discussed previously, although these designs share common elements, they also possess very basic differences. These differences are not just simple procedural or action variations but reflect distinct philosophical positions, research traditions, and orientations to data collection and analysis. In this text we could not possibly present and do justice to each of these orientations, their underlying foundations, and their thinking and action processes. We do, however, present to you some of the collective concerns and shared language of researchers along the naturalistic continuum. Then in Chapter 10 we present four designs that are used in health and human service research or that may be valuable to you as a researcher in the practice environment.

ELEMENTS OF THINKING PROCESSES

We have already discussed some of the basic characteristics of naturalistic inquiry. Let us review eight attributes and their implications for design.

Purpose

Designs along the continuum of naturalistic research vary in purpose from developing descriptive knowledge to evolving full-fledged theories about observed or experienced phenomena. As discussed in earlier chapters, when there is no theory to explain a human phenomenon or when the investigator believes that existing theory and explanations are not accurate, "true," or incomplete, naturalistic designs are preferred because their structure is exploratory, revealing new insights and understandings without the imposition of preconceived concepts, constructs, and principles. Although the specific purpose of a study may differ, all designs along the continuum of naturalistic inquiry seek to describe, understand, or interpret daily life experiences and structures from the perspective of those in the field.

Let us once again consider as an example Liebow's[1] ethnography of street corner life as presented in his classic book *Tally's Corner*. In his work, Liebow was interested in understanding the nature of a street corner culture to which he did not belong. Although there had already been extensive research on the population he chose to examine, these investigations had not characterized the group from the perspective of its members. Naturalistic method, in this case ethnography, was the only way to reveal the complexity and richness of the culture. Liebow gained valuable insights because his informants spoke for themselves and described their life experiences.

Context Specific

As in *Tally's Corner*, to discover "new truth" rather than impose an existing frame of reference, the investigator must go to the setting where the human phenomena occur or must ask individuals who experience the phenomena of interest about their perspectives. Because of the revelatory nature of naturalistic research, this form of investigatory action is conducted in a natural context or seeks explanations of the natural context from those who experience it from the inside. Naturalistic research is therefore context specific, and the knowing derived therefrom is embedded in that context and does not extend beyond it. The designs along the continuum of naturalistic inquiry share this basic attribute of context specificity and knowledge grounded in or linked to the data that emerge from a particular field.

Complexity and Pluralistic Perspective of Reality

Third, because of its underlying epistemology and its inductive approach to knowing, naturalistic research assumes a pluralistic perspective of "reality" that examines complexity.

Let us consider this point in more detail. One of the hallmarks of inductive reasoning, reasoning in which principles are extracted from seemingly unrelated information, is that the information can be organized differently by each individual who thinks about it. The end result of induction is the development of a complex relationship among smaller pieces of information, not the reduction of principles to their parts, as is the case of deductive reasoning. It is therefore possible that the same information may have different meanings or that there may be pluralistic interpretations to different individuals.

For example, in the study of prison violence by Maruyama,[2] the nature of violence was investigated in two separate institutional settings. The findings from each prison, although yielding some similarities, indicated cultural differences between the two prisons in the reported causes of violence and the inmates' interpretation of the meaning of that violence. The naturalistic design was able to reveal the pluralistic nature of prison violence while also shedding light on the basic elements that were shared in both settings.

Transferability of Findings

The findings from naturalistic design are specific to the research context only. That is to say, it is not the desired outcome of naturalistic design to generalize findings from a small sample to a larger group of persons with similar characteristics. So why bother doing it if you cannot use the research beyond the actual scope of the study? The answer lies in the purpose of naturalistic research as a theory generating tool. Naturalistic researchers use their methods and findings to generate theory and reveal the unique meanings of human experience within human environments. Because the investigator assumes that no current knowledge adequately explains the phenomenon under investigation, the outcome of naturalistic design is the emergence of explanation, principles, concepts, and theories. This point is illustrated by Liebow[1] in his claim:

There is no attempt here to describe any Negro men other than those with whom I was in direct immediate association. To what extent this descriptive and interpretive material is applicable to Negro street corner men elsewhere in the city or in other cities or to lower class men generally in this society or any other society, is a matter of further and later study. This is not to suggest that we are dealing with unique or even distinctive persons and relationships. Indeed, the weight of the evidence is the other direction [p. 14].

In this passage, even though the researcher is highlighting the contextual limitations of the knowledge, he also is indicating that the principles revealed in his study may have relevance to other arenas. Further, he suggests that additional research is the appropriate means to ascertain the degree of fit, or what Guba[3] would call the "transferability," of his findings to other similar populations and contexts.

Because phenomena are context bound and cannot be understood apart from the context in which they occur, naturalistic inquiry is not concerned with the issue of generalizability or external validity. Rather, the concern is with understanding the richness and depth within one context. The development of "thick[4]" in-depth descriptions and interpretations of different contexts lead to the ability to transfer meanings in one context to another. In so doing, the researcher is able to compare and contrast contexts and their elements to gain new insights as to the specific context itself.

Flexibility

In the experimental-type continuum, we said that design provides the basic structure and plan for the thinking and action processes of the research endeavor. However, in naturalistic research, design is a fluid, flexible concept. Design labels such as "ethnography," "life history," and "grounded theory" suggest a particular purpose of the inquiry and orientation of the investigator. However, these labels do not refer to the specification of step-by-step procedures, a blueprint for action, or a predetermined structure to the data gathering and analytic efforts. Rather, naturalistic designs are flexible.[5] It is expected that the procedures and plans for conducting naturalistic research will change as the research proceeds. For example, Willoughby and Keating[6] indicate in their study of caregivers of persons with Alzheimer's disease, "Ongoing field notes kept by the researchers assisted in the development of key concepts and in suggesting the direction of later interviews" [p. 33].

Not only do procedures change, but the nature of the research query, the scope of the study, and the manner by which information is obtained are constantly reformulated and realigned to fit the emerging "truths" as they are discovered and obtained in the field. This flexibility is illustrated more fully in our discussions of gathering information (see Chapter 15) and analytic actions (see Chapter 18).

Language

A major shared concern in naturalistic inquiry is understanding the language and its meanings for the people who are being studied. This concern with language is not just relevant to the ethnographer who studies a cultural context in which a language different from the investigator is used. Even within the same cultural or language context, people use and understand language differently. It is through language, symbols, and ways of expression that the investigator comes to understand and derive meaning within each context.[7] The investigator engages in a rigorous and active analytic process to translate, so to speak, the meaning and structure of the context of those studied into those meanings and language structures represented in the world of the investigator. The investigator must be careful to accurately represent the meanings and intent of expression in this translation or analytic and reporting process.

Emic and Etic Perspectives

Design structures vary as to their "emic" or "etic" orientation toward field experience. An "emic" perspective refers to the "insider's" or informant's way of understanding and interpreting experience. This perspective is phenomenologic in that experience is understood as only that which is perceived and expressed by informants in the field. Data gathering and analytic actions are designed to enable the investigator to bring forth and report the voices of individuals as they themselves speak and interpret their unique perceptions of their reality. The concern with the "emic" perspective is often the motivator for naturalistic inquiry as we discussed earlier for the concept of purpose.[8] For example, Krefting[9] explains why she decided to study head injury from an ethnographic perspective and, in doing so, highlights the preference for an emic viewpoint shared by designs along this continuum:

> Until recently, the investigation of health problems was dominated by the outsider perspective, in which important questions of etiology and treatment are identified by the medical profession. Studies based on this perspective assume that medical professionals are the authorities on what illness is and that they alone know what questions ought be asked. . . . .Ethnographic studies of the experience of illness and disability, . . . consider what an illness looks like from an insider's perspective [p. 68].

An "etic" orientation reflects the structural aspects of the field.[8] Many investigators integrate an emic and etic perspective. Typically field work will start by using an emic perspective or a focus on the voices of individuals. Then, further along the field work process, other pieces of information are collected and analyzed to place individual articulations and expressions within a social structural or systemic framework. As Krefting[9] explains, although she started with an emic consideration, she used a range of data collection strategies to integrate her understandings with an etic viewpoint. She[9] states that:

> While gathering data, I was able to review theory and check it for pertinence. There was, then, a constant movement back and forth between the concrete data and the social science concepts that helped explain them [p. 70].

Some designs, such as phenomenology and life history, favor only an emic orientation. Listen to how Frank[10] describes her preference for an emic perspective and the purpose of her life history approach:

> The life history of Diane DeVries represents a collaborative effort, between the subject and researcher, to produce a holistic, qualitative account that would bear on theoretical issues, but that primarily and essentially would convey a sense of the personal experience of severe congenital disability. The life history, conceived in this way, emerged from a humanistic interest in presenting the voices of people often unheard, yet whose lives were otherwise studied, and acted upon, based on data that are decontextualized and fragmented from the standpoint of the individual [p. 639].

Because the investigator enters the field having "bracketed," "suspended," or let go of any preconceived concepts, the naturalistic researcher defers to the informant or "experiencer" as the knower. This abrogation of power by the "investigator" to the "investigated" is characteristic of all naturalistic designs to a greater or lesser degree. An extreme example of relinquishing control over the research process is illustrated by Maruyama's[2] work. In the prison study, not only did the inmates function as the informants, but they planned and conducted the research in its entirety. We discuss a variation of this collaborative research effort in our discussion of participatory action research later in the book.

Where one stands regarding an emic or etic perspective shapes the overall design that is chosen and the specific data collection and analytic actions that emerge within the context of the field.

Gathering Information and Analysis

Finally, analysis in naturalistic designs relies heavily on qualitative data and is an ongoing process that is interspersed throughout data-gathering activities. Data gathering and analysis are interdependent processes. As one collects information or each piece of data, the investigator engages in an active analytic process. In turn, the ongoing analytic activity frames the scope and direction of further data collection efforts. This interactive, dynamic process is characteristic of all designs along the continuum of naturalistic inquiry and is explored more fully in Chapters 15 and 18.

SUMMARY

The purpose of naturalistic inquiry and the nature of the concept of design in this tradition are vastly different from the experimental-type continuum. The language and thinking processes that characterize naturalistic design are based on the notion that knowing is pluralistic and that knowledge derives from those who experience. Thus, thinking is inductive, the action processes are dynamic and changing, and they are carried out within the context where the phenomena of interest occur. The outcome of these thinking and action processes is the generation of theory, principles, or concepts that explain human experience in human environments and capture its complexity and uniqueness.

Within the category of naturalistic design, there are also variations in the way in which the investigator implements the eight essential characteristics described in this chapter. Designs vary from informant driven and unstructured to researcher driven and more structured. In Chapter 10 we place designs that are least structured and most informant driven at the bottom of the naturalistic continuum to denote that they create the foundation for more structured thought and action processes. These forms of inquiry are evolving their own standards, vocabulary, and criteria for design adequacy. These criteria are reviewed in both Chapter 15 and Chapter 18.

EXERCISES

1. Select a research article that uses naturalistic inquiry. Identify and provide evidence of the eight elements of the researcher's thinking process:

- Purpose of research
- Context of the research
- Pluralistic perspective of reality
- Concern with transferability
- Flexibility
- Concern with language
- Emic and etic perspectives
- Interactive/analytic process

2. Using the same article, determine some of the ethical issues that shaped the investigator's behavior in the field.

REFERENCES

1. Liebow E: *Tally's corner*, Boston, 1967, Little, Brown.
2. Maruyama M: Endogenous research: the prison project. In Reason P, Rowan J,

editors: *Human inquiry: a sourcebook for new paradigm research,* New York, 1981, John Wiley & Sons, pp 227–238, 267–282.

3. Guba EG: Criteria for assessing the trustworthiness of naturalistic inquiries, *Educ Commun Technol J* 29:75–92, 1981.
4. Geertz C: Thick description: toward an interpretative theory of culture. In *The interpretation of cultures: selected essays,* New York, 1973, Basic Books.
5. Marshall C, Rossman GB: *Designing qualitative research,* Newbury Park, Calif, 1989, Sage.
6. Willoughby J, Keating N: Being in control: the process of caring for a relative with Alzheimer's disease, *Qualitative Health Res* 1:27–50, 1991.
7. Hodge R, Kress G: *Social semiotics,* Ithaca, NY, 1988, Cornell University Press.
8. Headland TN, Pike KL, Harris M: *Emics and etics: the insider/outsider debate,* Newbury Park, Calif, 1990, Sage.
9. Krefting L: Reintegration into the community after head injury: the results of an ethnographic study, *Occup Ther J Res* 9:67–83, 1989.
10. Frank G: Life history model of adaptation to disability: the case of a congenital amputee, *Soc Sci Med* 19:639–645, 1984.

10

Continuum of Designs in Naturalistic Inquiry

- **Endogenous Research**
- **Critical Theory**
- **Phenomenology**
- **Heuristic Research**
- **Life History**
- **Ethnography**
- **Grounded Theory**
- **Summary**
- **Exercises**

- Grounded Theory
- Ethnography
- Life History
- Heuristic Design
- Phenomenology
- Critical Theory
- Endogenous Design

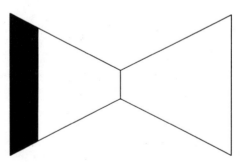

Using the language and thinking processes introduced in Chapter 9, we are ready to explore specific designs along the continuum of naturalistic inquiry. In this chapter we have selected varied designs to illustrate the different ways in which researchers along this continuum think about the research process and tackle the complexity of social experience.

As we indicated in Chapter 9, a shared principle of designs along the continuum of naturalistic inquiry is that phenomena occur or are embedded in a context, natural setting, or the "field." The term "fieldwork" refers to the basic activity that engages all investigators along this continuum. In conducting fieldwork, the investigator enters the natural setting without the intent of altering or manipulating conditions. The purpose is to observe, understand, and come to know so that one may describe, explain, and generate theory. However, how investigators come to know, describe, and explain phenomena differs among designs. Although each design described here shares the basic language and thinking processes explored in the previous chapter, they are based in distinct philosophical traditions. As such, these designs differ in respect to three basic aspects:

- Purpose of the research
- Use of the literature review
- Level of investigator involvement in the imposition of specific procedures and steps throughout the research process

In this chapter, we provide an overview of the basic framework of seven designs (endogenous research, critical theory, phenomenology, heuristic research, life history, ethnography, and grounded theory) and how they differ along the dimensions just listed. We discuss the actual action processes of doing naturalistic inquiry in Chapters 15 and 18.

ENDOGENOUS RESEARCH

Endogenous research represents the most open-ended, "nonstructured" approach to research along this continuum. In this approach, research is "conceptualized, designed and conducted by researchers who are insiders of the culture, using their own epistemology and their own structure of relevance." [p. 230].[1] The unique feature of this design-type is that the investigator relinquishes control of a research plan and its implementation to those who are the subjects of the inquiry. The subjects also are primary or co-investigators who work independently or with the investigator in determining the nature of the study and how it is to be shaped. It is the subjects and investigators who make decisions about and participate in the building and testing of information as it emerges.

Endogenous research can be organized in a variety of ways and may include the use of any research strategy and technique along both the naturalistic and experimental-type continua. What makes endogenous research a naturalistic design is its paradigmatic framework. That is, the investigator views knowledge as emerging from individuals who know the best way of obtaining that information. As Argyris and Schon[2] point out, this approach is based on a proposition developed by Kurt Lewin that:

Causal inferences about the behavior of human beings are more likely to be valid and enactable when the human beings in question participate in building and testing them. Hence it aims at creating an environment in which participants give and get valid information, make free and informed choices (including the choice to participate), and generate internal commitment to the results of their inquiry [p. 86].

This design is placed at the base of the naturalistic continuum because the research query is turned over to the "subjects." It gives "knowing power" exclusively to the persons who are the subject of the inquiry and is characterized by the absence of any predefined "truths" or structures. Because endogenous research is conducted by insiders, there are no external principles that guide the selection of thought and action processes within the study itself.

Endogenous research has also been referred to as "action research" and more recently "participatory action research" (PAR).[3] This latter term has emerged from research on organizations. Its purpose is to close the gap between research and implementation of new organizational practices by involving multiple interest groups concerned with the study problem. The PAR approach, as in endogenous research, involves an ongoing exchange between researcher and study participants in the diagnosis and evaluation of problems and in the data gathering process and assessment of findings.[4]

Let us look at an example of endogenous research on prison violence by Maruyama.[5] Maruyama entered the prison environment as a collaborator and participant observer rather than as the research consultant. This is an important characteristic of endogenous research. As a collaborator, the researcher participates as an equal member on a team rather than as a consultant to recommend or dictate procedures and direction. Based on a series of group meetings, two teams comprised of prisoners who had no formal education in research methods created purposes and plans of action that were important and meaningful to each group. Maruyama[5] describes the research in this way:

A team of endogenous researchers was formed in each of the two prisons. The overall objective of the project was to study interpersonal physical violence (fights) in the prison culture, with as little contamination as possible from academic theories and methodologies. The details of the research were left to be developed by the inmate researchers [p. 17].

The prisoners chose to use both experimental-type interview techniques and a form of qualitative data analysis to characterize the violence in the environments in which they were insiders. Maruyama found that the prisoners not only were able to collect extensive and meaningful data but also were quite capable of conducting sophisticated conceptual analyses of their data set. The findings from a study "owned" and conducted by insiders yields very different results from one in which a researcher enters the prison environment as a stranger to ask questions regarding the reasons and nature of violence. From this study, violence was explicated in the language of the inmates, and the findings clearly displayed the cultural nature of violence in each prison. Just think of how different the findings could have been if an outsider went to a group of inmates and asked questions about why prison violence occurred. The endogenous methodology not only gave the inmates a voice but revealed insights that could never have been uncovered by techniques in which the investigator took the authoritative position in the project.

The concerns of the investigator in this type of research include how to effectively build and work with a team, how to relinquish control over process and outcome, and how to shape the group process to move the research team along in the study of themselves.

CRITICAL THEORY

As implied by the name, critical theory is not a research method but a world view that suggests both an epistemology and a purpose for conducting research. Whether critical theory is a philosophical, political, or sociologic school of thought has been debated in the contemporary literature. In essence, critical theory is a response to postenlightenment philosophies, positivism in particular, which "deconstructs" the notion of a unitary truth that can be known by one way or method. That is to say, critical theory is a movement, which we suggest lives in the world of philosophy, and views the world through a sociopolitical lens.

Critical theorists seek to understand human experience as a means to change the world. The purpose of research founded in critical theory is, therefore, to come to know about social justice and human experience as a means to promote social change.

Critical theory was born in the Social Institute in the Frankfurt School in the 1920s. As the Nazis gained power in Germany, critical theorists moved to Columbia University and further developed their notions of power and justice. With their focus on social change, critical theorists came to view knowledge as power and the production of knowledge as "socially and historically determined" [p. 6].[6] Derived from this view is an epistemology that upheld pluralism or in which coming to know about phenomena occurs in multiple ways. Furthermore, knowing is dynamic, changing, and embedded within the sociopolitical context of the times. According to critical theorists, there is no one objective reality that can be uncovered through systematic investigation.

Critical theory shares other principles in naturalistic inquiry such as a view of informant as knower, the dynamic and qualitative nature of knowing, and a complex and pluralistic world view. Furthermore, critical theorists suggest that research crosses disciplinary boundaries and challenges current knowledge generated by experimental-type methods. Because of the radical view posited by critical theorists, literature on the research process is used primarily as a means to understand the status quo. Thus, the action process of literature review may occur before the research, but the theory derived therefore is criticized and deconstructed, taken apart to its core assumptions. The hallmark of critical theory is, however, its purpose of social change and empowerment. Critical theory relies heavily on interview and observation as methods by which data are obtained. Strategies of qualitative data analysis, discussed in detail in Chapter 18, are the primary analytic tools used in critical research agendas.

Consider this example. Suppose you were interested in understanding the influence of the relationship between client and formal service provider on the outcome of therapeutic intervention with clients who are substance abusers. First, the researcher using a critical theory perspective would critically review the literature from the assumption that a power imbalance exists between client and practitioner. Furthermore, this imbalance has the effect of disempowering the client. Second, naturalistic inquiry strategies would be used to understand the relationship from the perspectives of both parties. Third, the understanding derived from the research would be used to promote social change.[7]

Because critical theory has the specific aim of social change and empowerment of the subject of inquiry, we have placed it just above endogenous method on the naturalistic continuum.

PHENOMENOLOGY

The specific focus of phenomenologic research is investigation of everyday experience and an interpretation of the meaning of those experiences. However, different from many other types of inquiry in which the researcher himself or herself attributes meaning to experience in the analytic phases of the inquiry, phenomenologists believe that meaning can be understood only by those who experience it. Thus, phenomenologists do not impose an interpretive framework on data but look for it to emerge from the information that they obtain from their informants. Phenomenologic research is further anchored on the notion of the limitations of methods to share and communicate experience.[8] "The phenomenon we study is ostensibly the presence of the other, but it can only be the way in which the experience of the other is made available to us" [p. 4].[8] Methods of obtaining information rely on biography. However, different from life history, the informant himself or herself interjects the primary interpretation and analysis of experience into the interview. In phenomenologic research, the action process of literature review is framed with the phenomenologic principle of the limits of communication. Thus, the literature may be used to illustrate the constraints of our understanding of human experience or may be used to corroborate the communication of the other.

For example, suppose you were interested in investigating the experience of aging. As a phenomenologic researcher, you would recruit an informant and then engage in lengthy discourse with that informant about his or her life experience and meanings that the informant attributed to his or her lived experience. You might turn to the literature to ascertain the commonalties of what is communicated to other researchers and your informant, or you may use the literature merely to suggest a rationale for using phenomenologic methodology without relying on substantive theory to inform your study.

We placed phenomenologic research low on the naturalistic continuum because of the researcher's dependence on those investigated and their willingness and ability to express and reveal their experiences into the structure of the inquiry and the interpretation of the data.[9, 10]

HEURISTIC RESEARCH

Heuristic research is another important design type along the naturalistic continuum. According to Moustakas[11] heuristic research is "a research approach which encourages an individual to discover, and methods which enable him to investigate further by himself" [p. 207].

Heuristic research is a design strategy that involves complete emersion of the investigator into the phenomenon of interest and self-reflection of the investigator's personal experiences. The investigator engages in intensive observation and listening to individuals who experience the phenomenon and logs their individual experiences as well. The meanings of those experiences are then interpreted and reported. The premise of this approach is that knowledge emerges from personal experience and is revealed or known to the investigator through his or her own experience of the phenomenon.

Moustakas[11] provides an example of heuristic research. When his daughter became ill, he experienced his own loneliness and realized that health care professionals revealed, by their behavior, that they did not seem to understand the nature of loneliness in persons who were sick. Moustakas, therefore, engaged in an extensive project where he examined his own loneliness, listened to the stories of hospitalized children, and then examined the literature on loneliness to further reveal the meaning of the concept. In this design, the literature served as another form of data. It was not used as a source for defining the concept of loneliness before the field was entered.[11]

As you can see from this example, heuristic research is conducted by an individual for the purpose of discovery and understanding of the meaning of human experience. The research involves total emersion in the experience of humans as a way of coming to understand their perspectives. The experiences of the researcher and information derived from a literature review are considered primary and critical sources of data as well. Note the function of and timing in which a literature review is conducted. The information that emerges from all of these sources take the form of narrative, which is then analyzed (see Chapter 18).

Heuristic design is more structured than endogenous research in the direction the study follows. The researcher has a dominant role in collecting and analyzing information. In this design, the investigator's own feelings and experiences become the object of study and a source of data. Its very name, "heuristic," suggests the essence of this design that it serves as a foundation for further inquiry into the human experience that it describes.

LIFE HISTORY

Life history design is placed on the naturalistic continuum because of its focus on individual lives in the context of the social environment. It is placed lower on the continuum than ethnography because the investigator looks primarily to the informant not only to produce the data but also to provide analysis of their meaning. As described by the name, life history research aims to reveal the nature of the "life process traversed over time."[12] Because of the recognition that individual lives are unique, researchers using life history methodology focus their research on one individual at a time, even though investigators may vary in scope and purpose for conducting the research. For example, some investigators may be searching for universals that characterize human experience as discovered through the examination of multiple individual lives, whereas others may be interested only in discovering the essence of an individual's lived experience. Depending on the aim of inquiry, life history researchers use literature in variable ways. Investigators who are attempting to analyze the value of theory in explaining the complexity of a human life may use the literature as an organizing framework, whereas researchers who are seeking new theoretical or unique understandings may use the literature at different junctures in the research thinking and action processes.

In life history methodology, the sequence of life events and the meaning of those events to the unfolding of a life are examined from the perspective of the informant.[13] Studies seek to uncover and characterize marker events, or "turnings,"[14] defined as specific occurrences that shape and change the direction of individual lives. Typically researchers rely heavily on unstructured interview, which may begin with statements such as "tell me about your life history" and then narrow the field of inquiry to the examination of meanings of the events. Participatory and nonparticipatory observation are also useful information gathering strategies.

Consider an example from Frank's[14] research. Frank was interested in enhancing the understanding of disability. She chose life history methodology as the method that could best characterize the unique meanings of disability to the life course of an individual with an amputated limb. Through in-depth interview, Frank was able to uncover not only the chronology of life events but also the symbolic and practical meaning of those events to her informant. Analysis of turnings allowed Frank to examine the type of experiences that were most influential in determining the future direction of the informant's life.

Life history studies can be retrospective or prospective. In health and human service research, the majority of life history studies are retrospective in their approach. Informants are asked to reconstruct their lives and to reflect on meaning. Because it can be costly and extremely time consuming, prospective longitudinal studies are often beyond the resources of the researcher.

ETHNOGRAPHY

Ethnography is a method derived primarily from the discipline of anthropology. Although there are many schools of thought that present different thinking and action processes of ethnography,[15-20] we present the basic elements that ethnographers tend to have in common.

Ethnography is a naturalistic design in which the intent is to understand the underlying patterns of behavior and meaning of a culture. Because there are so many definitions of culture, we suggest the broadest synthesis of those that are most common. In this book we define culture as the set of explicit and tacit rules, symbols, and rituals that guide patterns of human behavior within a group. The ethnographer as an "outsider" to the cultural scene seeks to obtain an "insider" perspective. Through extended observation, emersion, and participation in the culture, the ethnographer seeks to discover and understand rules of behavior.[15] Data are collected through several primary methods: (1) interview and observation of those who are willing to inform the researcher about behavioral norms and their meanings, (2) the researcher's participation in the culture, and (3) examination of meaning of cultural objects and symbols. Insiders who willingly engage with the investigator are called "informants." Informants are the investigator's "finger on the pulse of the culture," without which the investigator would not be able to achieve full understanding. Although the results of ethnography are specific to the culture being studied, some ethnographers attempt to contribute to theory of universal human experience through this important naturalistic methodology.

An ethnography begins with gaining access to a culture. The investigator then characterizes the context, or "social scene,"[15] by observing the environment in which the culture operates. Equipped with an understanding of the cultural context, the ethnographer uses participant observation, interview, and examination of materials or artifact review to obtain data. Through field notes, audio recordings, video recordings, or combination thereof, qualitative data are logged.[21] Analysis of the data is ongoing and moves from description to explanation, to revealing meaning, to generation of theory. Impressions and findings are verified with the insiders and then reported once the findings are determined to accurately represent the culture. Reflexivity analysis,[22, 23] or the analysis of the extent to which the researcher influences the results of the study, is an active component of the research process. These processes are described in more detail in Chapters 15 and 18.

We have classified ethnography as a naturalistic design because of its reliance on qualitative data collection and analysis, the assumption that the researcher is not the knower, and the absence of a priori theory, or a theory imposed prior to entering the field. We place ethnography further up the naturalistic continuum because of the degree of control that the investigator purposely retains over the research process. It is the investigator who makes the decisions regarding the who, what, when, and where of each observation and interview experience. The view that knowledge can be generated about the "other" without the viewpoint of the investigator influencing the study is a different philosophical approach from heuristic and endogenous designs. Furthermore, ethnography is a design that is capable of moving beyond description to reveal complex relationships and theory.

In health and human service research, ethnography or research based on ethnographic principles[24] is becoming a widely accepted research approach. Investigators use ethnography to obtain an "insider's" perspective on the meaning of health and social issues as a basis from which to develop meaningful health care and social service interventions. Ethnography has also been used to examine various service environments. One example of such a study is the ethnography of the nursing home environment conducted by Savishinsky.[24] Several strategies, including interview and participant observation, were used to describe the culture of the nursing home and to understand the meaning of life in that culture. Through an analysis and synthesis of the perspectives of residents and staff, Savishinsky was able to point to directions for change in that environment.

GROUNDED THEORY

Grounded theory is defined as "the systematic discovery of theory from the data of social research" [p. 3].[20] It is a more structured and investigator directed strategy than the previous designs along the continuum of naturalistic inquiry that we have thus far discussed. Its purpose is primarily to develop and verify theory.

Developed primarily by Glaser and Strauss,[20] it represents the integration of a quantitative and qualitative perspective in thinking and action processes. The primary purpose of this design strategy is to evolve a theory or, as the name implies, ground a theory in the context in which the phenomenon under study occurs. The theory that emerges then is intimately linked to each datum of daily life experience that it seeks to explain.

This strategy is similar to the designs we have already discussed in its use of an inductive process to derive concepts, constructs, relationships, and principles to understand and explain a phenomenon. It is distinguished from other naturalistic designs by its use of a structured data gathering and analytic process termed the *constant comparative method.* In this approach, each datum is compared with others to determine similarities and differences. Glaser and Strauss, in numerous books and articles, along with others, have developed a very elaborate scheme by which to code, analyze, recode, and produce a theory from narrative obtained through a range of data collection strategies.

Let us briefly consider an example of the constant comparative method. In a study of homeless middle-aged women, Butler[25] used constant comparison in combination with other strategies to develop her analysis. Initially she examined the data set to induce categories of data that were repeated throughout the experiences of the women who served as her informants. As she analyzed each datum, either an experience, articulation, or observation, she compared it to the data in the existing categories to determine similarities and differences between new and previously obtained information. If it fit, the datum was coded with an existing code. If not, a new category or subcategory was developed.

The purpose of the comparative method is not only to reveal categories but also to explore the diversity of experience within categories, as well as to identify links among categories.

Glaser and Strauss also suggest that grounded theory strategies are capable not only of generating theory but of verifying it. A query using a grounded theory approach begins with broad descriptive interests and through data collection and analysis moves to discover and verify relationships and principles.

SUMMARY

In this chapter we examined seven designs and have shown how they differ in their structure, purpose, use of the literature, and the extent to which the researcher shapes or structures the project design. Endogenous research has the least amount of structure built in by the researcher, does not use literature to guide the investigation initially, and relies most heavily on informants as knowers and interpreters of that knowing. In designs further up on the continuum, the investigator more directly guides the thinking and action processes and relies on a literature base at different junctures in the research process.

With this introduction to some of the more frequently used designs on the naturalistic continuum, you may want to explore each in more depth. We refer you to the bibliography for sources that address each design, their use, and conduct.

EXERCISES

1. To understand the different purposes of designs along the continuum of naturalistic inquiry, formulate at least four distinct research queries that lead to four different designs discussed in this chapter.

2. Go to a public place and determine how you would conduct a study using ethnography to determine public behavior patterns. Then determine how you would conduct the study using heuristic research.

3. Plan how you would use a naturalistic methodology to discover the health beliefs of a chronically ill older population living in the community. Now plan a study with children who are chronically ill and living with their families. How would your strategies differ?

REFERENCES

1. Maruyama M: Endogenous research: rationale. In Reason P, Rowan J, editors: *Human inquiry: a new paradigm sourcebook*, New York, 1981, John Wiley & Sons.
2. Argyris C, Schon DA: Participatory action research and action science compared: a commentary. In Whyte WF, editor: *Participatory action research*, Newbury Park, Calif, 1991, Sage, pp 85–96.
3. Whyte WF, editor: *Participatory action research*, Newbury Park, Calif, 1991, Sage.
4. French W, Bell CH Jr: *Organization development, behavioral sciences interventions for organization improvement*, Englewood Cliffs, NJ, 1984, Prentice-Hall.
5. Maruyama M: Endogenous research: the prision project. In Reason P, Rowan J, editors: *Human inquiry: a sourcebook of new paradigm research*, New York, 1981, John Wiley & Sons, pp 267–282.
6. Tierney W: *Culture and ideology in higher education*, New York, 1991, Praeger.
7. Nencel L, Pels P: *Constructing knowledge: authority and critique in social science*, Newbury, Calif, 1991, Sage.
8. Darroch V, Silvers RJ: *Interpretive human studies: an introduction to phenomenological research*, Washington, DC, 1982, University of American Press.
9. Douglas J, editor: *Understanding everyday life*, London, 1970, Routelage and Kegan Paul.
10. Van Manen M: *Researching lived experience: human science for an action sensitive pedagogy*, Albany, NY, 1990, State University of New York Press.
11. Moustakas C: Heuristic research. In Reason P, Rowan J, editors: *Human inquiry: a sourcebook of new paradigm research*, New York, 1981, John Wiley & Sons, pp 207–218.
12. Rabin AI, Zucker RA, Emmons RA, et al: *Studying persons and lives*, New York, 1990, New York, Springer-Verlag, p IX.

13. Langness LL, Frank G: *Lives: an anthropological approach to biography,* Novato, Calif, 1981, Chandler and Sharp.
14. Frank, G: Life history model of adaptation to disability: the case of a congenital amputee, *Soc Sci Med* 19:639–645, 1984.
15. Spradley JP: *Participant observation,* New York, 1980, Holt, Rinehart & Winston.
16. Sperber D: *On anthropological knowledge,* Cambridge, Mass, 1987, Cambridge University Press.
17. Levi-Strauss C [Jacobson C, Schoepf BG, translators]: *Structural anthropology,* New York, 1960, Basic Books.
18. Agar M: *Speaking of ethnography,* Newbury Park, Calif, 1986, Sage.
19. Geertz C: *The interpretation of cultures: selected essays,* New York, 1973, Basic Books.
20. Glaser B, Strauss A: *The discovery of grounded theory,* New York, 1967, Aldine.
21. Lofland J, Lofland L: *Analyzing social settings: a guide to qualitative observation and analysis,* ed 2, Belmont, Calif, 1984, Wadsworth.
22. Ruby J: Exposing yourself: reflexivity, anthropology and film, *Semiotica* 30:153–179, 1980.
23. Steier F, editor: Research and reflexivity, Newbury, Calif, 1991, Sage.
24. Savishinsky JS: *The ends of time: life and work in a nursing home,* New York, 1991, Bergen & Garvey.
25. Butler S: *Homeless women,* unpublished dissertation, Seattle, University of Washington, 1991.

11

Integrating Methods, Designs, and Purposes

- **Meaning of Integration**
- **Continuum of Integration**
- **Integrated Design Strategies**
- **Summary**
- **Exercises**

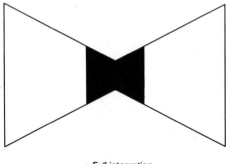

- Full integration
- Mixed method
- Triangulation

By now we hope you have an understanding and appreciation of the distinct ways in which researchers think along the naturalistic and experimental-type continua. In this chapter, we explore ways in which designs and thinking processes can be combined. The different ways of combining studies fall along a continuum we refer to as the *integrated continuum of designs*. First let us discuss the meaning of integration and the continuum it represents, and then we shall present two design strategies, case study and focus group, which use different levels of integration.

MEANING OF INTEGRATION

An "integrated" approach has become increasingly important for health and human service research. The intent of integration is to strengthen a study by selecting and combining designs and methods from both paradigms so that one complements the other to benefit or contribute to an understanding of the whole.

As you will recall, we have emphasized throughout this book that there are many different thinking and action processes from which to choose to answer a systematic inquiry. We have stressed the need to evaluate the particular strengths and weaknesses of any one method or design strategy in relationship to an investigator's specific research purpose and research query or question. Integration is based on an investigator's firm understanding of the strengths and limitations of thinking and action processes within each paradigm, design strategy, or particular methodologic approach. Based on this knowledge, the investigator can mix and combine strategies to strengthen the research effort and to strive for comprehensive understanding of a phenomenon under consideration. Brewer and Hunter[1] use the term "multimethod" research and explain the premise of this approach to inquiry in this way:

> Our individual methods may be flawed, but fortunately the flaws in each are not identical. A diversity of imperfection allows us to combine methods not only to gain their individual strengths but also to compensate for their particular faults and limitations. . . Its (multimethod research) fundamental strategy is to *attack a research problem with an arsenal of methods that have nonoverlapping weaknesses in addition to their complementary strengths* [p. 17].

Although scholars disagree about the extent to which paradigms can be combined, there is a growing movement on both sides of the research continua to at least recognize the strengths and values of the other.[2, 3] Experimental methodologists are beginning to recognize that naturalistic inquiry represents a legitimate and logical alternative research strategy, whereas qualitative researchers are giving greater attention to standardization of analysis and procedures and to the complementary role of experimental-type designs.

The attempt to integrate research paradigms as a way to strengthen and advance scientific inquiry has close to a 30-year history in sociology, anthropology, and education but has only more recently become a focus in health and human service research. The movement toward the integration of designs in health and human service research has been motivated by four factors.

First, the focus on the experimental-type continuum as the only viable research option has been brought into question by findings that repeatedly indicate little to no measurable effects of experimental treatments in health and human service research.[4, 5]

Second, there is growing recognition that many of the persistent and unanswered issues that are critical to health and human service cannot be answered by the experimental paradigm.[6]

Third, there is an increasing realization that quantitative information, in and of itself, does not necessarily provide insight as to the various processes that underlie the "hard facts" of a numeric report.[3]

Fourth, there are often great discrepancies in knowledge generated from qualitative and quantitative studies.[7] In earlier chapters we demonstrated how each thinking and action process is rooted in a philosophical tradition and how each design strategy may generate knowledge that is different from that produced by another design strategy. Remember the discussions of research on moral development in men and women and the vast body of often conflicting literature on caregiving? These discrepancies in the literature present both an intellectual and practical dilemma. This dilemma is particularly acute in health and human services where research is often interpreted in regard to its immediate consequences for implementation and practice.

These four points of tension in the research world have led many investigators to question the traditional opposition of qualitative-quantitative research and to suggest strategies to overcome or transcend the limitations of each design strategy.[8]

In this chapter we do not intend to solve the historical debate and philosophical dilemma inherent in the attempt to combine distinct epistemologies and traditions. However, we do believe that on a practical level, each design continuum presents thinking and action processes that serve as tools of the trade of a researcher. Paradigms are ways to study aspects of the world, and as we emphasize in this book, they must be selected based on a fit between the assumptions of the paradigm and its tools and the nature of the phenomenon under study. By understanding the intent of these tools, we can *purposely* and *systematically* combine and integrate diversity to enhance and strengthen an inquiry. The combination of different perspectives to examine the same phenomenon has the potential of producing a more holistic and comprehensive understanding than the use of singular strategies.

Integrated research is not really a new approach in research. Many researchers have implicitly used one or more of the forms of integration we discuss in this chapter. For example, there is a long history of the use of simple descriptive statistics such as frequencies and means in ethnographic research. Or quantitative studies frequently use an exploratory, open-ended series of questioning before establishing closed-ended and numeric based interview questions. What is new today is the purposeful and logical development of an integrated perspective that combines elements of the thinking and action processes from both the naturalistic and experimental-type continua at each step of a research project. These integrated studies transcend the paradigm debate and enhance scientific inquiry.

CONTINUUM OF INTEGRATION

We suggest that an integrated perspective can be conceived as a continuum that ranges in complexity of design structure. The most basic form of integration, and one that most researchers are familiar with, is *triangulation*. The next level of integration is often referred to as mixed methods or multimethod approaches, whereas the most complex level of integration involves the use of multiple paradigms to answer a single broad research inquiry. The way in which methods are combined and an integrated perspective is used reflects one's paradigmatic framework and viewpoint of the value and role of designs. Table 11–1 summarizes each level of integration, paradigmatic orientation, and specific strategies. Let us examine each form of integration to illustrate this point.

Triangulation

The most basic form of integration has been referred to as triangulation. Although in Chapters 15 and 18 we discuss a particular form of triangulation used in data collection and for an analytic strategy, here we discuss the broad meaning of the concept.

There are many definitions of triangulation in the literature. The term was first used in the social sciences to refer to the use of multiple methods to measure a single construct.[9] It has since been expanded to include other forms such as theoretical, data, and investigator triangulation.[10] In each type of triangulation, different techniques are linked to counteract the weakness of the other and provide information on a single phenomenon. The purpose of this type of triangulation is to confirm information regarding a single phenomenon and obtain what is called "convergent validity." Convergent validity of a finding is achieved by bringing together information collected, tested, or analyzed through more than one method, what Campbell and others have referred to as "multioperationalism." This form of triangulation is used primarily by research-

TABLE 11–1.

Summary of Integrated Strategies

Level	Paradigmatic Orientation	Strategies
Triangulation	Naturalistic inquiry Experimental type	Completeness function Convergence function
Mixed methods	Either continuum Experimental type Either continuum	Nested Traditional sequential model Parallel model
Full integration	Complementarity	Integration of purpose, problem formulation, data collection, analysis, reporting conclusions

ers along the experimental-type of continuum and has been referred to as the "convergence function" of triangulation.[11] In this approach the emphasis is on the confirmation of discrete constructs and the refinement of measurement approaches.

An investigator can triangulate within or across methodologic strategies. An example of triangulation within one methodologic approach involves the use of different scaling techniques in a single survey instrument to measure the same phenomenon. Each scale provides confirmation of the reliability of the information that is derived. If the scales are truly measuring a similar phenomenon, the investigator would achieve convergence across the measures. An example of triangulation across methodologic approaches may involve the use of chart extraction, open-ended interviews, and a structured questionnaire to see if there is convergence in the findings across methods that have different purposes and strategies.

Triangulation has been used somewhat differently by researchers involved in naturalistic inquiry. Along this continuum of designs, triangulation has been referred to as a multistrategy approach. The purpose of this approach is to combine different methods to reveal an additional piece of the puzzle or to uncover varied dimensions of one phenomenon.[11] Referred to as the "completeness function," different methods are purposely chosen because each assesses a different aspect of the dimension of the problem under study. For example, an investigator might combine open-ended interviewing with direct observation and chart extraction to achieve a complete understanding of recovery issues associated with a cerebrovascular accident. Each data collection strategy provides an understanding of the total experience of recovery.

As you can see, in a triangulated approach to integration, the researcher's thinking and action processes reflect a particular paradigm. How one uses triangulation, as either a "completeness function" or "convergent function," depends on the purpose of the investigation and the investigator's paradigmatic or epistemologic framework. With either of these functions, multiple strategies are combined from within that paradigm or across paradigms for the purpose of either achieving "completeness" or "convergence."

Mixed Methods

On the second level along the continuum of integration are strategies that have been referred to as mixed methods. There are three basic strategies. In one approach, which we refer to as a "nested" strategy, the investigator develops a research inquiry based on his or her paradigmatic framework and then borrows specific methodologic techniques from another paradigm or design strategy in an effort to strengthen an aspect of the study. An example is an ethnography that uses probability sampling at some point in the field to identify a representative sample. The ethnographer then conducts a survey on a specific issue that emerged during data collection.

Smith and Louis[5] identify two other basic approaches to mixing methods: sequential and parallel. In the traditional *sequential approach*, the investigator views naturalistic inquiry as a descriptive, exploratory form of research that precedes the primary study that is quantitative. Thus, qualitative methods such as in-depth interviewing may be used in the construction of an instrument, for example, or to identify dimensions that should be considered in a quantified study. A newer sequential approach uses designs along the continuum of naturalistic inquiry to follow up on quantitative results to explain, elaborate, or interpret findings. For example, case material may be collected after a survey or another standardized data collection approach to illuminate the questions that arise in an analysis of a numeric data base.

The other approach to integration at the mixed method level is referred to as the *parallel model*. This model makes no assumptions about a linear relationship between qualitative and quantitative methods. In this "separate but equal" approach,[3] designs from the two continua are perceived as representing two very different ways of knowing that help the researcher see and illuminate different aspects of a phenomenon under study.

Full Integration

This level represents the most complex form of integration. It involves an inquiry that integrates multiple purposes and thus combines design strategies from different paradigms so that each contributes knowledge to the study of a single problem to derive a more complete understanding of the phenomenon under study. Paradigms along the continuum of naturalistic inquiry can be combined (a hermeneutic approach with grounded theory, for example, in a study by Wilson and Hutchinson[12]), or designs from the continuum of naturalistic inquiry can be combined with designs from the continuum of the experimental type.

The innovative feature of this perspective is that designs along both continua are perceived as equal but separate and as such can complement each other. Neither a quantitative nor qualitative approach takes precedence over the other, but both are used cojointly to derive a comprehensive understanding.[13] This vision of "complementarity"[8] is perhaps the most controversial form of integration and the one most debated in the literature.

At this level along the integrated continuum, integration occurs at each step of the research process. As Brewer and Hunter[1] point out, integration of different and often conflicting empirical findings that stem from studies using diverse methods typically occurs after the fact of the research. It occurs at a highly abstract level through the synthesis of published studies and the development of theory to provide an "umbrella" for diverse ideas that appear in the literature. Although integration of research through literature review is critical to advance knowledge and the research enterprise[14] as we discussed in Chapter 5, this traditional approach is limited and cannot solve gaps and contradictions across studies. As Brewer and Hunter[1] comment, integration at the project level, however, "begins the task of integration from the ground up by calling upon individual social scientists to integrate methods throughout the course of their individual investigations. In multimethod research, one must confront diversity and try to resolve contradictions from the outset" [p 24].

When a fully integrated study is conducted, there are three research challenges. First, one needs to state a research problem so that it warrants an integrated approach. Second, one must systematically develop ways of comparing and evaluating the results of different methods. Third, one must reconcile contradictions as they emerge in data collection and analysis. As more researchers participate in integrated designs, models will be developed to logically link diverse methods and causal and interpretive analyses.

Let us examine an example of a complex integrated study. Bluebond-Langner,[15] a medical anthropologist and her colleagues, a group of sociologists and social workers, wanted to understand the role of camps designed for children with cancer. They were specifically interested in examining how the camp experience enhanced the children's knowledge of cancer and its treatment. They[15] describe the purposes of their study as "to determine (1) whether cancer and its treatment are discussed informally among the children, (2) what kinds of information are exchanged if such discussions take place, and (3) how these interactions might affect the children's knowledge and understanding of cancer and its treatment" [p. 207].

As you can see, each of these purposes is embedded in a distinct research paradigm and set of assumptions, but all are combined to strengthen an understanding of the camp experience on children's knowledge. For example, one intent of the study was to understand the *process* of knowledge transmission in the camp or how children came to know and understand cancer and its treatment. This process orientation fit with the underlying assumptions of the investigators that knowledge emerges from and is transmitted through interactions among campers and between counselor and camper. This perspective lent itself to an ethnographic-type of design to discover what types of knowledge were transmitted and how learning occurred in the absence of formal camp instruction.

A second intent of the study was to examine the extent of knowledge acquired through camp and how this experience affected the children's knowledge level. The underlying question here was whether the camp experience had a positive impact and increased the children's level of understanding about the disease process and its treatment. Obviously it was not feasible to use a two-group randomized design in which children were either assigned to cancer camp or stayed home to test the impact of camp on knowledge and understanding. Nor was that solely the intent of the study, although perhaps many cancer associations and camp administrators would have liked to have been able to demonstrate effectiveness. Instead, the investigators developed a quasi-experimental design, akin to a time series approach, where multiple measures to tap knowledge attainment were obtained before camp, during camp (the experimental intervention), and after the experience. In this way, the investigators were able to demonstrate change in level of knowledge over time.

Bluebond-Langner et al.[15] describe their design in this way:

> A case-study, quasiexperimental design was used for this study. The design combined extensive participant observation, qualitative interviewing and objective measurement of the children's knowledge before and after the camp experience. This design enables an investigator to document and describe the presence of particular phenomena within a specific group (in this case, the camper's discussions and knowledge about cancer and its treatment). It also allows the investigator to generate hypotheses to explain the existence of and changes in these phenomena [p. 208].

Through statistical analyses of repeated measurements, the investigators were able to demonstrate an increase in the knowledge of cancer treatment over time obtained by children. Through ethnographic observations and interviews, the investigators were able to provide an explanation of how participation in camp led to increased knowledge of cancer treatment. Interestingly, it was through the insights gained by the qualitative analytic component of the study that the investigators were able to suggest that alternative competing hypotheses as to why children demonstrated increased knowledge were not adequate. Thus, in the absence of the ability to develop a control group, the ethnographic component yielded critical insights to support the inference that it was the camp experience and the exchanges among campers that produced greater understandings among the children. This is an example of the complementary role of two design strategies and of how each aspect of the research process—from problem formulation, data collection, to analysis—was integrated to achieve completeness in understanding of a single phenomenon.

INTEGRATED DESIGN STRATEGIES

Let us turn to a discussion of two design types that represent the continuum of integration. These two design strategies, case study and focus group, are used by researchers working from various traditions along both the experimental-type and naturalistic continua. Each design can be used in single or multisite studies in health and human service. They each combine a range of data collection, sampling, and analytic strategies. A number of worthwhile books listed in the bibliography provide the details of each one of these research approaches. Our purpose in this chapter is to present an overview of the thinking processes of alternative designs as examples of integrated perspectives.

Case Study Design

The case study is a detailed, in-depth description of a single unit, subject, or event. Often referred to as an "n of 1" type of study, it is longitudinal in nature in that data are collected over a period of time. However, it can also use historical materials and a retrospective strategy.

Historically, case study has been considered an inferior approach to generating knowledge because of its focus on a single unit of analysis. However, in the past 2 decades, case study has evolved into a well-respected set of design strategies and has been used with increasing frequency in health and social science inquiry.

Although there are diverse definitions of case study, Yin's[16] perspective captures its breadth and purpose:

> A case study is an empirical inquiry that investigates a contemporary phenomenon within its real-life context; when the boundaries between phenomenon and context are not clearly evident; and in which multiple sources of evidence are used [p. 23].

Many different design strategies are possible in a case study approach. These designs span the two research continua from experimental to endogenous research. Case study designs may yield quantitative or qualitative data, may be limited to one case or involve many, and may explore, describe, or predict. A case may be defined as a single contemporary phenomenon that embodies multiple parts and can be examined using different sets of information. The most well-known example of a "case" is an individual. However, there are many other types of cases that can be the focus of an investigation, including organizations, neighborhoods, single events, hospitals, outpatient departments, or specific groups that are examined as a whole. The key to understanding the concept of "case" lies in how the research question or query is framed.

When to Select Case Study Design

When do you select case study design? As mentioned earlier, case study is appropriate when (1) you are interested in examining a phenomenon in its current context; (2) you want to contribute to support for and development of theory; and (3) you want to explore in-depth a case that is atypical, different, or in which it would be impossible to conduct a group design.

However, one of the most important factors that influences the selection of case study is that of practicality. In health and human service research, an investigator frequently has questions about a phenomenon but is unable to conduct a large scale study. Case study fits well into the daily routine of health and human service providers. In a case study, you are not necessarily concerned with generalizing findings from a sample to a population. Rather, the findings that you encounter can be used to suggest theory or to provide empirical support for existing theory. In other words, case study results are not generalizable from samples to populations but may be used to generalize to theory. The case study approach provides additional data for the development or support of a multifaceted understanding of a phenomenon.

Let us consider an example. Suppose you are a physical therapist who is conducting a work tolerance program with persons with low back pain. Each patient brings unique concerns to the program. You are interested in examining your program to determine what individual characteristics seem to fit best with which program components. You might consider studying each individual as a case to understand how individual and programmatic factors mix to result in certain outcomes. You already have your data set: the initial assessments and your set of progress notes. How convenient! You do not have to set up experimental conditions or deviate far from your usual procedures to conduct your study. This example brings us to another very important use of case study research: evaluation of practice.

Although evaluation has become a field unto its own, research design is the foundation of evaluation. Practitioners, funders, and policy makers must know "what works" in this complex health and social service system. Many designs are used, but case study design is very popular and appropriate to evaluation concerns. Let us consider why. First, in evaluation of practice, you are not concerned directly with how a study of the outcome and process of your practice generalizes to other practices. Therefore, your unit of focus is one phenomenon—your practice. Second, case study, where a client is treated as a single unit of analysis, provides information from which to examine the process of therapeutic practice and its outcomes. Third, organizations and programs can also be treated as one unit of analysis, rendering the case study approach very desirable.

There are two major considerations in the selection of a case study design. Because in a case study we do not select "representatives" of a population as our sample, we cannot generalize from case study findings to a population. Second, the use of group statistical data analytic techniques to aggregate and analyze the cases involved in each separate study is debatable because we do not use "sampling logic" to select the cases.

Types of Case Study Design

Yin[16] suggests four specific case study designs: holistic single case study, embedded single case study, holistic multiple case study, and embedded multiple case study. Knowing the case study design classifications as Yin poses them provides guidelines for selecting the number of cases and the type of case to be examined. Let us briefly examine each.

Holistic case studies are those in which the unit of analysis is seen as only one global phenomenon. In embedded case studies, the cases are conglomerates of multiple subparts or are subparts themselves placed within larger contexts. For example, if we were to select an organization as a case, a holistic approach would examine the function of the entire organization as one unit. On the other hand, the embedded case study would examine the parts of the organization and relate the function of each of the parts to the function of the whole.

In single case study designs, only one study on a single unit of analysis is conducted. In multiple case studies, more than one study on single units of analysis are conducted. The single holistic case study is conducted only once on one case. The multiple holistic case study examines a global unit of analysis more than once in more than one case. A single embedded case study focuses on multiple parts of a single case using only one case, whereas a multiple embedded case examines more than one case containing many subparts.

When thinking about holistic versus embedded designs, the investigator must consider the nature of the phenomenon that is of interest. Does the unit of analysis have naturally occurring parts that will reveal relevant information? Or does the "whole" provide the most informative approach?

Let us consider the example of family violence. The investigator who believes that family violence is a cultural phenomenon may select a holistic case study design to analyze the response of the total family to cultural sanctions for violence. On the other hand, an embedded case study would be the design of choice for the investigator who is attempting to determine the dynamics among individuals that could potentially provoke family violence. In the holistic approach, the family would be observed and tested as a whole. In the embedded approach, individual members of the family would be the subparts of the single family case.

What about selection of single vs. multiple case study? If replication is the purpose, a multiple case study approach is most appropriate. However, if an investigator wants to understand a unique case or examine a unique embedded phenomenon, single case study is appropriate. For example, suppose you are a nurse who is treating clients with acquired immunodeficiency syndrome (AIDS) in their homes. In your client group, you find one individual who seems unusual in her ability to cope with terminal illness. To understand that individual further and to discover the factors that contribute to coping strategies, you conduct a single case study. This "deviant" case study approach can illuminate distinctions in coping strategies. However, if you are interested in characterizing the experience of persons with AIDS in their communities, you might be more apt to select a multiple case study design. About now you may be asking why the investigator would select a multiple case study over another type of group design. The answer lies in the definition of case study design. Case study, in this and other situations, is ideal for describing persons over time in their contemporary contexts without sacrificing the complexity of human experience.

Data analysis in case study design is controversial. Although many consider it inappropriate to use group techniques to analyze case study data, other researchers disagree. However, we do suggest that analytic strategies such as descriptive statistics, visual presentation of changes in quantitative measures, and inductive analyses of narrative information are useful analytic tools in case study designs.

Focus Group Methodology

Focus group methodology, originally a market research approach to understanding consumer behavior and preferences, is being used with increasing frequency in health and human service.[17, 18] Focus groups tend to be cost effective, are relatively easy to conduct, and are a quick way to obtain rich information regarding a particular phenomenon. The approach is especially suited to research questions that seek to understand or uncover such processes as how health professionals make decisions or how individuals experience different disease processes or disability. A focus group may involve anywhere from 5 to 15 individuals who are chosen to participate in a 1- to 2-hour group session. The meeting is usually led by a moderator who is skilled in group processes. The session can either be structured, semistructured, or unstructured. The group interview approach facilitates an interactive process to uncover the ways in which participants think and act regarding a particular preidentified phenomenon.

It is usually recommended that a minimum of three to five focus groups for any one theme or topic be conducted to achieve saturation in the information gathered. As Morgan[19] and others[17] have suggested, if the moderator can guess what a participant will say, a point of saturation has been achieved and no further significant information is usually obtained thereafter. After a focus group session, it is good practice to debrief with participants and review initial impressions of general patterns or themes that emerged from the group session with co-investigators. This debriefing can also be tape recorded and analyzed. The purpose of the debriefing is for the investigators to compare notes and begin to summarize and interpret the key issues which immediately present themselves. This process is characteristic of other qualitative methodologies in that the analysis of materials is ongoing and influences the conduct of the next focus group. For example, interpretations from one focus group can be confirmed and refined in a subsequent group. In addition, interesting insights or conflicting stories can be identified by the investigators and then pursued in the next focus group session. Thus, the analytic process is often incremental and builds from one group to the next. The direct involvement of the investigator and other members of a research team provide consistency in interpretations and a mechanism to confirm emerging interpretations. The final analytic strategy may involve both a qualitative, thematic approach and then a content analysis in which specific use of words or response patterns are counted.

This type of design can be part of a mixed methods approach in either a naturalistic type of inquiry or with an experimental-type design. Traditionally focus groups have been used in a sequential model. In this approach, the focus groups are initially used to generate descriptive information to identify the parameters of a phenomenon. Based on the focus group information, a structured questionnaire is developed and an experimental-type study design such as a survey is implemented to substantiate, refine, or extend the findings from the focus groups. More recently, focus groups have been used to explain variations or relationships which are identified in a survey or other quantitative-type design. In this newer sequential approach, a series of focus groups are systematically planned to provide explanation to a quantitative finding.

Within the structure of a focus group, a range of strategies from both the experimental type and naturalistic continua are often mixed or combined. For example, although convenient or purposeful sampling is usually used to obtain participants, one could also target volunteers through a systematic sampling plan and sampling frame to increase the representativeness of participants. Furthermore, a key to focus group interviewing is that the questions are both deductive, based on current knowledge about the topic, and inductive, which follow up on insights that emerge during the interview. Thus, interview formats tend to be open and only partially specified to allow the exploration and discovery of new material. This approach is particularly suited for an initial study to uncover the dimensions and boundaries of a phenomenon. But focus groups are also used to provide insights into patterns that have been identified by survey research. Some focus groups use very structured, fixed questions to purposely explore predetermined areas of interest. The degree of structure depends on the research purpose and the nature of the research question and participants.

SUMMARY

In this chapter we have presented three levels of integration: triangulation, multimethod and full integration. In each level we have discussed some of the major strategies that have been developed to achieve an integrated perspective. An integrated perspective may not be appropriate for every study. First, multimethod research can be costly and time consuming because additional project tasks are implemented. Also, investigators working by themselves may not have the necessary expertise to combine strategies from different paradigms. However, integration at its most basic form, that of triangulation, should be considered in any study as a way of strengthening a design.

Integration, although not necessarily a new research perspective, has attracted a new following of researchers who are actively working to build logical connections among research paradigms. Full integration, based on the principle of complementarity, offers the opportunity to examine health and human service research inquiries in the context in which they occur and without compromising the complexity of the phenomenon under study.

EXERCISES

1. Select a research article from the *Journal of Qualitative Health Research* that uses triangulation as a "completeness function." Identify the strategies used and how each adds to the understanding of the other.
2. Select an experimental-type research article. Redesign the study by using an integrated approach.
   a. In what ways does your redesign strengthen the research?
   b. What are the potential limitations of your redesign?

REFERENCES

1. Brewer J, Hunter A: *Multimethod research,* Newbury Park, Calif, 1989, Sage.
2. Pearlin L: Structure and meaning in medical sociology, *J Health Soc Behav* 33:1–9, 1992.
3. Rowles GD, Reinharz S: Qualitative gerontology: themes and challenges. In Reinharz S, Rowles GD, editors: *Qualitative gerontology,* New York, 1988, Springer-Verlag, pp 3–33.
4. Ottenbacher K: *Evaluating clinical change: Strategies for occupational and physical therapists,* Baltimore, 1986, Williams & Wilkins.
5. Smith AG, Louis KS, editors: Multi-method policy research: issues and applications, *Am Behav Sci* 26:1–144, 1982.
6. Evolving methodology in disability, *Rehabil Brief* XII:5, 1989.
7. Mechanic D: Medical sociology: some tension among theory, method and substance, *J Health Soc Behav* 30:147–160, 1989.
8. Salomon G: Transcending the qualitative-quantitative debate: the analytic and systematic approaches to educational research, *Educ Researcher* 20:10–18, 1991.
9. Campbell D, Fiske D: Convergent and discriminant validation by the multi-trait, multi-method matrix, *Psychol Bull* 56:81–104, 1959.
10. Duffy M: Methodological triangulation: a vehicle for merging quantitative and qualitative research methods, *Image J Nurs Scholarship* 19:130–133, 1987.
11. Morse JM: *Qualitative nursing research,* Rockville, Md, 1989, Aspen.
12. Wilson HS, Hutchinson SA: Triangulation of qualitative methods: heidiggerian hermeneutics and grounded theory, *Qualitative Health Res* 1:263–276, 1991.
13. Rossman GB, Wilson BL: Numbers and words: combining quantitative and qualitative methods in a single large scale evaluation study, *Evaluation Rev* 9:627–643, 1985.
14. Cooper HM: *Integrating research: a guide for literature reviews,* ed 2, Newbury Park, Calif, 1989, Sage.

15. Bluebond-Langner M, Perkel D, Goertzel T, et al: Children's knowledge of cancer and its treatment: impact of an oncology camp experience, *J Pediatr* 116:207–213, 1990.

16. Yin RK: *Case study research,* Newbury Park, Calif, 1989, Sage.

17. Patton MQ: *Qualitative evaluation and research methods,* ed 2, Newbury Park, Calif, 1990, Sage.

18. Kreuger R: *Focus groups: a practical guide for applied research,* Newbury Park, Calif, 1988, Sage.

19. Morgan DL: *Focus group as qualitative research,* series no 16, Newbury Park, Calif, 1988, Sage.

Part III

ACTION PROCESSES OF RESEARCH

Let us reexamine what we have discussed thus far. In Part I we introduced the importance of research for the helping professional and how research differs from other ways of knowing about the world. Then in Part II we showed how the level of theory development in one's area of interest, as evidenced in a literature search, and one's philosophical framework influence the formulation of a research question and query. We also explored traditional design classifications and the distinct language and thinking processes used by researchers along the two continua in the development of a design strategy. Finally, we explored ways in which design strategies can be integrated.

Now that you have grasped the vocabulary, emphases, and general thinking processes of researchers along both continua, we are ready for action. In the chapters that follow in Part III we focus on how researchers implement or put into action the four aspects of a design:

1. Bounding your study and obtaining individuals, concepts, or other phenomena to study (see Chapter 12)
2. Collecting information or data using a range of strategies (see Chapters 13 through 15)
3. Analyzing and interpreting information or data (see Chapters 16 through 18)
4. Reporting and using information and conclusions (see Chapter 19)

In our discussions we constantly refer to five factors that have a major impact on the selection of an action process:

- Philosophical or epistemologic framework of the researcher
- Investigator's specific research purpose

- Nature of the research question or query
- Type of design
- Resources available and practical limitations or challenges of the environment in which the research occurs

Remember, both thinking and action processes are planned within an epistemologic, purposeful, and theoretical context.

Each chapter in Part III introduces specific research action processes and how they differ in naturalistic inquiry and experimental-type research. You will learn diverse and varied ways to implement and put into action the setting of boundaries, data collection, and data analytic and data reporting strategies to best respond to the problems you wish to answer.

12 | Setting Boundaries: Sampling and Domain Selection

- **Boundary Setting for Experimental-Type Designs**
- **Boundary Setting for Naturalistic Designs**
- **Subjects, Respondents, Informants, or Participants?**
- **Summary**
- **Exercises**

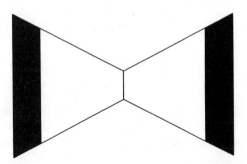

DOMAIN SETTING
- On-going boundary setting process
- Judgement selection of informants
- Convenience selection of informants
- Selection of setting(s)

POPULATION DEFINITION
- A priori definition of population characteristics
- Probability sampling
- Non-probability sampling

Once a research problem and an appropriate design have been developed, you are ready to consider how individuals, concepts, or objects will be selected for your study. This selection process involves setting boundaries or limitations as to what or who will be included and excluded from your study.

Setting limitations or boundaries as to what and who is studied occurs in each and every design type regardless of its placement on the continua. One important way researchers set boundaries is by limiting the scope of the investigation to a specified group of individuals, phenomena, or conceptual dimensions. For example, if you are interested in surveying the needs of parents with children with chronic disabilities, it would be impossible from a monetary and management standpoint to interview every person who fit into that category in the United States. Or let us say you are interested in the historical development of the profession of physical therapy. It would not be feasible, time-wise, to examine every historical detail or written document to understand the context of professional growth. In each of these studies, you would need to make decisions as to which parents to survey or which documents to examine. Limiting the number of parents with children with disabilities is an example of setting boundaries by limiting the *persons* who will be studied. Deciding which historical documents to examine is an example of limiting your study by specifying the boundaries of the *concept* in which you are interested.

There are numerous ways to limit the scope of a study. As in each of the previously discussed thinking processes of research, an investigator chooses a method to bound a study based on five considerations:

1. Researcher's philosophical framework
2. Purpose
3. Nature of the question
4. Research design
5. Practical considerations regarding access to the object or objects of inquiry

In experimental-type designs, boundaries tend to be set through two processes: population definition and sampling. In naturalistic designs, boundary setting is a process that occurs throughout the total research process. However, it tends to be accomplished by defining a domain and the point access from which to enter that domain. The inquiry then describes the natural context in which the inquiry unfolds. This domain may be geographical, as in the selection of a community in which to study health perceptions, or the domain may be a group of individuals, such as the selection of persons with traumatic brain injury living in the community.[1] Then, within the context of that boundary, selection of persons to interview, scenes or events to observe, or materials or artifacts to review unfolds and occurs on an ongoing basis as the researcher becomes increasingly more focused. These selections can be made using a variety of approaches and may include those traditionally associated with experimental-type research.

In this chapter we first discuss the ways in which researchers set boundaries for experimental-type designs and then discuss how boundary decisions are made in naturalistic studies. As you will see, each continuum approaches boundaries and boundary setting quite differently.

BOUNDARY SETTING FOR EXPERIMENTAL-TYPE DESIGNS

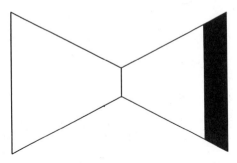

SAMPLING PROCESS

In experimental-type design, boundary setting is deductive. That is, the researcher begins with an idea of who he or she wishes to study, clearly defines the characteristics of the group or *population* that includes the individuals to be studied, and then selects a subset or *sample* from that group to participate in the actual study. The group of interest to the investigator is called a population. A population is defined as a group of persons, elements, or both that share a set of common characteristics that are defined by the investigator. A sample is defined as the subset of the population that participates in or is included in the study. The process of selecting a sample is called *sampling.*

The main purpose of sampling is to be able to accurately draw conclusions about the population by studying the smaller group of elements, the sample. The problem faced by the experimental-type researcher is how to best select a part of a population so that the part can most accurately represent the whole. Accurate representation is critical for the findings from the sample to be generalized to the larger group from which the sample is drawn.

Sampling designs or procedures have been developed to increase the chances of selecting individuals or elements that will be most representative of the larger population from which they are drawn. The more representative the sample, the more assured the researcher is that the findings from the sample also apply to the population. The extent to which findings from a sample apply to the population is called *external validity*. Most researchers conducting experimental-type designs try to maximize external validity by using one of the sampling procedures discussed here and summarized in Table 12–1.

Sampling Process

Investigators follow many different procedures to draw a sample from the population. The first step in sampling involves the careful definition of a

TABLE 12–1.

Summary of Commonly Used Sampling Plans

Probability Sampling	Nonprobability Sampling
Parameters of a population are known	Parameters of a population are not known
Use of sampling frame	No sampling frame available
Every member/element has same probability of being selected for sample	Probability of selection not known
Examples:	Examples:
Simple Random Sampling Use of table of random numbers to randomly select sample	*Convenience Sampling* Available individuals enter study
Systematic Sampling Determine sampling interval width and select individuals	*Purposive Sampling* Deliberate selection of individuals for study
Stratified Random Sampling Randomly select subjects from predetermined strata that correlate with variables in study	*Snowball/Network Sampling* Informants provide names of others who meet study criteria
Cluster Sampling Successive random sampling of units to obtain sample	*Quota Sampling* Attempt to include individuals unlikely to be represented

population. As we indicated earlier, a population is defined as the complete set of elements that share a common set of characteristics. Examples of elements are persons, households, communities, hospitals, or outpatient settings. Regardless of the type of element, an element is the *unit of analysis* included in the population. For example, in a study of the management practices of hospitals, the investigator is interested in the hospital as a whole rather than the people within it. The hospital is the unit of analysis, and the sample would be comprised of a specified number of hospitals. In a study of the characteristics of patients admitted to hospitals in the summer months, each patient is the unit of analysis, and the sample would include individual persons.

The investigator defines or chooses the characteristics or parameters of the elements of a population that are important in a study. Let us say we are interested in studying the needs of parents caring for children with chronic illness. Our first step is to identify the specific characteristics of the population we are interested in examining. Suppose we decide to *include* parents caring for children between the ages of 3 and 8 years who have been diagnosed with cystic fibrosis and are currently hospitalized. We also might decide to *exclude* cases in which both parents do not reside together, are both unemployed, and in which there is more than one other sibling in the home. By establishing these inclusion and exclusion criteria, we have defined a *target population* from the universe of parents caring for a child with a chronic illness. In some instances, the target population is an ideal group because such a population cannot be accessed by the investigator. However, in this chapter, we define the target population as the group of individuals or elements from which the investigator is able to select a sample.

Each element of the target population may possess other characteristics than those specified by the investigator. However, each element:

1. Must possess all of the characteristics that the investigator has identified as inclusion criteria
2. Must not possess any of the characteristics that the investigator has defined as exclusion criteria
3. Must be available, at least in theory, for selection into the sample

How do investigators identify a population? First, in experimental-type designs, the research question itself guides the investigator in identifying a population. Second, the researcher uses the literature to further clarify and provide support for establishing population parameters. For example, if an investigator is interested in persons with chronic mental illness, there is a vast body of literature that provides clear descriptions and definitions of this broad category of individuals. An astute investigator may choose a well-accepted source such as the DSM-III-R for a definition of chronic mental illness.[2]

The literature also helps to identify those characteristics that the investigator may want to exclude from the targeted population. In the same example, persons might be excluded from the target population if they possess a primary physical diagnosis along with a psychiatric diagnosis.

Let us consider another example. Suppose you are interested in understanding the daily routines of persons with quadriplegia. The first step would be to define the population characteristics of the group you would like to study. Some guiding questions you would ask yourself are:

1. Should I include both males and females in the study?
2. Should all persons with quadriplegia be included regardless of functional level?
3. In what type of setting should persons included in the study reside? In the community? In an institutional setting?
4. In which geographic location should they live?
5. What ages should be included?
6. What family styles should be included?
7. Should persons with additional diagnoses be excluded?

Your answers to these questions would lead to the development of inclusion and exclusion criteria. To answer these questions, you would first reflect back on your initial purpose in conducting the study and your research question. Then you would refer to the literature for definitions such as "quadriplegia" and "functional level."

Defining populations is an important step in sampling and must be done carefully and thoughtfully. Santiago et al.[3] illustrate how two different ways of defining a target population can lead to distinct findings. One study[3] of the homeless mentally ill population specified that a person was homeless "if residing on the streets, in automobiles, or in shelters or missions, at the time of their entrance into the study" [p. 1100]. A second study extended the inclusion criterion to those individuals who lived in a household in the past 3 months but who had previously met the homeless criteria for the first study. By opening up the selection criteria, the investigators obtained significant differences in the gender balance between the two studies.

The second step in boundary setting in experimental-type design involves drawing a sample through the use of a *sampling plan*. There are two basic categories of sampling plans: probability sampling and nonprobability sampling. Table 12–1 provides an overview of the commonly used sampling plans we discuss in this chapter.

Probability Sampling

Let us first discuss probability sampling. Probability sampling refers to those plans that are based on probability theory. The two basic principles of probability theory as applied to sampling are that (1) the parameters of the population are known, and (2) every member or element has an equal probability or chance of being selected for the sample.

The important point is that the probability of each element being included is known and greater than zero. By knowing the population parameters and the degree of chance that each element may be selected, an investigator can calculate the *sampling error*. Sampling error refers to the difference between the values obtained from the sample and the values that actually exist in the population. It reflects the degree to which a sample is representative of the population. The larger the sampling error, the less representative the sample is of the population and the more limited is the external validity. The purpose of probability sampling is to reduce sampling error and increase the external validity of a study.

Sampling error may be caused by either random error or systematic bias. Random error refers to those errors that occur by chance. Not much can be done about random error at the sampling stage of the research process. On the other hand, systematic error or bias reflects a flaw in the sampling process in which scores of subjects differ systematically from the population.

Sampling plans based on probability theory are designed to minimize this cause of error.

To use probability sampling, the investigator must be able to develop a *sampling frame* from which individuals or elements are selected. A sampling frame is defined as a listing of *every* element in the target population. Examples of sampling frames include a telephone book, the yellow pages, a complete list of hospitals in a specific region, and any other such lists. Defining a sampling frame can be a simple or a complicated process. A simple approach involves defining the sampling frame as those population members whom the investigator can easily access. A more complex approach involves developing a sampling frame to ensure that the total population can be accessed by the investigator. This approach involves more time and money. For example, in a study of reasons for attrition from occupational therapy, Bailey[4] developed her sample from two sampling frames. She used a roster of respondents who were members of the American Occupational Therapy Association (AOTA) and who had indicated in a member survey that they were not working. She also used a roster of occupational therapists who had allowed their membership to the professional association to lapse.

Because only AOTA members were considered, the sampling frame was composed of only those occupational therapists with a membership. Thus, the target population for Bailey's study was occupational therapists who were AOTA members but who were not working in OT positions. Figure 12–1 illustrates the relationship among population, target population, sampling frame, and sample.

There are four basic sampling procedures that can be used to select individuals or elements from a sampling frame: simple random sampling (SRS), systematic sampling, stratified random sampling, and cluster sampling.

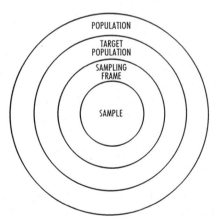

FIG 12–1.
Levels in sampling process.

Simple Random Sampling

Simple random sampling is the simplest method used to enhance representativeness of a sample. The term random does not mean haphazard. Rather, it means that theoretically each and every element in the population has an equal chance of being included in the sample. If elements are chosen by chance, they also have an equal chance, theoretically, to be exposed to all conditions to which all other members in the population can be exposed by chance. This random nature of selection, therefore, precludes the possibility of the sample being selected because of a special trait that is uncommon in the target population or of being exposed to an influence that does not theoretically affect the total population. Returning to Bailey's[4] study, a sample was randomly selected from the two rosters provided by AOTA, assuring that each element (individual) who met the criteria for inclusion had an equal chance of being selected for the study.

Most researchers who want to attain a large sample size use a table of random numbers to determine which elements should be included in the sample:

Subset From Table of Random Numbers

345	687	798	143	203	107	654	176
241	267	925	007	115	003	409	115
076	861	623	351	090	065	190	325

These tables, only a portion of which is displayed above, are very carefully constructed through computer generated programs to ensure a random listing of digits that appear with the same frequency.

This is the way the table is used. Let us say you plan to survey all full-time nonsupervisory nurses in a selected hospital (target population). You would obtain a complete listing of nurses who fit the study criteria (the sampling frame). Then you would assign each nurse on the list a unique number. If there were 150 nurses, the list would run from 001 to 150. Because the numbers contain three digits, you would select the random numbers in sets of three. To use the table, you begin by randomly selecting a starting point and then establishing a plan as to how to move through the table. Let us say that we decide to start from the third column from the left and one row down (925) and proceed to read down each row. Numbers outside the range of the numbers assigned to the nurses on the list (for example, numbers greater than 150) are ignored. Therefore, the first two numbers, 925 and 623, are outside the range and do not yield a selection. Moving to the top of the fourth row, numbers 143 and 007 do reflect number assignments of nurses in the sampling frame. We would thus proceed through the table as such until the total number of nurses needed for the study had been selected.

Random sampling can be done with or without replacement. Using a table of random numbers involves replacement because it is possible to select a number more than once. Random sampling without replacement can be as simple as drawing sample member names from a hat that contains all the names listed on a sampling frame. Let us consider what might happen with smaller numbers in the sampling frame when random sampling without replacement is used. Suppose we have ten persons in our sampling frame and we wish to select three persons for our sample. Before any subject is selected, the chances are 1 in 10 that an individual will be chosen. However, once that name is chosen and not replaced in the pool, the next individual has a 1 in 9 chance of being selected. As you can see, the rule of equal chance for selection is violated. However, if we replace the name of the subject who was selected into the hat for the subsequent two selections, the chances for sample selection remain 1 in 10. Replacement assures the same probability of chance of selection for each element throughout the selection process.

Systematic Sampling

Simple random sampling can often be time consuming and difficult to complete when one is drawing a very large sample. Systematic sampling offers a simpler method to randomly select elements. It involves determining a sampling interval width based on the needed sample size and then selecting every Kth element from a sampling frame. For example, let us say you want to survey a sample of 200 health care professionals from a total population of 1000. First, a sampling fraction (the interval width), 200:1,000, or 1 in 5, is derived. Second, a random number between 1 and 5 would be selected, such as 3, to determine the first person in the sampling frame to be selected. Then every fifth person from the starting point on the list would be selected for inclusion in the sample (the 8th, 13th, 18th person, and so on). Or a number from the table of random numbers is used to determine the starting point. If the sampling fraction is not a whole number, the decimal is usually rounded upward to the next largest whole number. When this sampling approach is used, it is critical that the sampling frame represents a random listing of names or elements and that there are no hidden biases, purpose, or cyclical arrangements to the way in which elements are arranged. For example, suppose that just by chance every tenth health professional on the list had just entered the practicing arena. If we used a sampling interval of 5 and had started with the fifth person, we would have selected only novice practitioners. This would have introduced systematic bias in our sample and increased the sampling error.

Stratified Random Sampling

Systematic sampling and SRS treat the target population as a whole. In stratified random sampling, the population is divided into smaller subgroups, or *strata.* Elements are then chosen from each strata. This sampling plan is a more complex approach to sampling. It enhances sample representation and lowers sampling error on a number of predetermined characteristics by making each subgroup more homogeneous and less variable. The more homogeneous a population, the fewer the number of elements needed to enhance representation and the lower the error in generalizing from part to whole. The Nielsen polls are an example of complex definition of sampling frames and samples. Here the researchers clearly specify a multiplicity of characteristics that typify the television watching population in the United States. Because this population is so diverse, it is separated into strata based on such characteristics as geographic location, socioeconomic status, age, and gender. The proportion of each subgroup, or *stratum*, in the total population is then determined. Sample subjects are drawn from each stratum in the same proportions that are represented in the population. Proportionate stratified sampling is frequently used when it is known that a given characteristic appears disproportionately or unevenly in the population. If we used SRS or systematic sampling, we might not obtain a sample that represents the proportions of various characteristics as they exist in the population.

Let us consider how a researcher may use this technique. Suppose in our population of persons with quadriplegia, only 20% are female. The researcher would split the sampling frame into two strata representing each gender. Then 20% of the sample from the female group and 80% from the male group would be selected. This selection process would ensure that gender is proportionately represented in the chosen sample. Without stratification, it is likely that the gender balance of the sample would not match that in the population. Characteristics are chosen for stratification based on the assumption that they will have some effect on the variables under study. In the previous example, gender may actually relate to daily routine. Thus, it would be important to ensure that such a comparison between males and females could be made with regard to daily activity. In Bailey's[4] study of attrition in occupational therapy, the sampling frame was stratified into three groups: members of AOTA who were not working in OT positions but were interested in returning to the field, members of AOTA who were not working in OT positions but were not interested in returning to the field, and persons who had allowed their membership to AOTA to expire. By stratifying the sampling frame, Bailey was able to examine differences for attrition that were related to the variable of "intention" to return.

Cluster Sampling

Cluster sampling, also referred to as multistage or area sampling, is another subject selection plan that involves a successive series of random sampling of units. Here the investigator begins with large units, or clusters, in which smaller sampling units are contained. This technique allows the investigator to draw a random sample without a complete listing of each individual or unit. For example, in testing a rehabilitation intervention for persons with a cerebrovascular accident in inpatient rehabilitation settings, the researcher would first list the geographic regions of the United States. Using simple random selection procedures discussed earlier, one would select geographic regions. Within each selected region, the investigator would obtain a complete listing of the free-standing rehabilitation hospitals and make a random selection among these units. Then within each selected hospital, the investigator would randomly select from a list of individual patients.

Nonprobability Methods

In experimental-type design, probability sampling plans are preferred to other ways of obtaining a sample when the goal is to generalize from sample to population or, in other words, to increase the external validity of the study. Randomization provides a degree of assurance that selected members will represent those who are not selected. These methods are often used in health and human service research for needs assessments, survey designs, or large-scale funded research projects. However, random methods are often impractical because using them requires the researcher to have considerable knowledge about the characteristics and size of the population, access to a sampling frame, access to a large number of elements, and the ability to ethically omit elements from the study. In practice, various forms of *nonprobability sampling* techniques are effectively used for experimental-type designs.

In nonprobability sampling, nonrandom methods are used to obtain a sample. These methods are used when the parameters of the population are not known or when it is not feasible or ethical to develop a sampling frame. Let us examine four basic nonprobability sampling methods.

Convenience Sampling

Convenience sampling, also referred to as accidental, volunteer, or opportunistic sampling, involves the enrollment of available subjects as they enter the study until the desired sample size is reached. The investigator establishes inclusion and exclusion criteria and selects those individuals who volunteer to participate in the study. Examples of convenience sampling include interviewing individuals in a doctor's office or enrolling subjects as they enter an outpatient clinic.

Purposive Sampling

Purposive sampling, also called judgmental, involves the deliberate selection of individuals by the researcher based on certain predefined criteria. In the example of the study of persons with quadriplegia, we might decide to purposely choose only those who are college educated and are able to articulate their daily experiences. In this way we purposely select individuals to represent insight into the daily routines of a larger group. Or we might ask clinical staff which of their patients with quadriplegia are "representative," or typical, of the clients in their site.

Snowball Sampling

Snowball sampling, or networking, involves asking subjects themselves to provide the names of others who may meet study criteria. This type of sampling is often used when researchers do not have access to a population. For example, in their study of how women addicts cope with the issue of AIDS, Suffet and Lifshitz[5] used a snowball sample. They asked the counselors in the agency from which they selected their sample to "nominate" [p 55] informants who, in turn, were also asked to identify additional informants. In this way, Suffet and Lifshitz were able to access a population that otherwise would have been difficult to find.

Quota Sampling

This technique is often used in market research. The goal of quota sampling is to obtain different proportions of subject types that may be underrepresented by using convenience or purposive sampling methods. In quota sampling, parameters of a population and their distribution in the population are known. The researcher then purposively selects a sample that is representative of the population in that elements are selected to display parameters in the same proportions exhibited in the population.

Let us consider an example. Suppose you are interested in testing the effects of a model day-care intervention on the functional level of persons with schizophrenia living in the community. Because schizophrenia represents different types of syndromes, you first determine the proportions of varying types of schizophrenia in your population. You then find that gender and age are distributed differently in each type. To assure that your nonrandom sample is as representative of your population as possible, you would set up a matrix of diagnoses, ages, and genders and select your sample by filling each cell in the same proportions that are exhibited in your population.

Summary

Experimental-type designs increase external validity by achieving representativeness of the sample through probability sampling. However, in health and human service research, probability sampling may not only be impractical but may be unethical. If you are unable to use probability sampling in your research, does that mean that your project is flawed? Not at all. The key to using nonprobability sampling is to attain the greatest degree of representation as possible and to clearly identify to your readers, the limitations of your findings. For example, in a study of attitudes of allied health students toward persons with disabilities, Estes et al.[6] selected a sample from two universities using a nonrandom process. In their discussion, they indicate that the generalization of findings to other student groups needs to be substantiated by further study in which larger sample sizes and additional university settings are tested. They clearly identified their sample selection procedures and, in the discussion, point out the application of the findings, as well as the limitations, on generalization and implications for further replication.

Comparison of Sample and Population

We started out by saying that the major reason for sampling is to assure representativeness and the ability to generalize from sample findings to the target population. The power of probability sampling is that population estimates can be made based on statistical theory. Five steps are involved in determining if findings from the sample are representative of that population:

1. State hypothesis.
2. Select level of significance.
3. Compute calculated statistical value.
4. Obtain a critical value.
5. Reject or fail to reject null hypothesis.

These steps are briefly reviewed here so that you can become familiar with this process.

The first step involves a statement of the hypothesis of no difference (null hypothesis) between the population and sample being compared. This null hypothesis implies that the values obtained in the sample, let us say, on a depression scale, are the sample values we would have obtained had we measured depression in other samples drawn from the same population. Second, the researcher selects a level of significance or the probability that defines how rare or unlikely the sample data must be before the researcher can reject the null hypothesis of no difference. If the significance level equals 0.05, the researcher is 95% confident that the null hypothesis should be rejected. Rejecting the null hypothesis means that there is a difference between the sample and the population. Third, using a statistical formula, a statistical value is computed. Fourth, the statistical value is compared with a critical value. Most statistical texts have tables of critical values to which you can refer. The critical value indicates how high the sample statistic must be at a given level of significance to reject the null hypothesis. If the sample value is greater than the critical value, the null hypothesis can be rejected, and the researcher concludes that a significant difference exists between sample and population. In many instances, the researcher wants to accept the null hypothesis, as in the case when he or she wants to demonstrate that sample values mirror or represent the larger group from which it was selected. In other cases, the researcher wants to reject a hypothesis of no difference when he or she wants to demonstrate that a particular intervention changed the sample significantly from the values represented in the population.

Determining Sample Size

Determining the number of participants in the study or the sample size is a critical issue and one with which new investigators often have difficulty. A common suggestion you may receive is to obtain as many subjects as you can afford. However, a large sample size is not always the best policy and is often unnecessary. The size of one's sample can influence the type of data collection techniques used, procedures for recruitment, and the costs involved in conducting the study. Although very few published research articles discuss the rationale for the size of the sample for a study, you should have a reason for the number you chose. A defense of your choice of a sample size is particularly important when you submit a proposal to a funding agency requesting funds to conduct the study.

The number of elements in a sample is determined once the population and sampling frame have been identified. The determination of sample size may be based on the proportion of units from the population that are necessary to conduct statistical testing. The number of subjects needed to test a hypothesis is directly related to the issue of what is called *statistical power*. This term refers to the probability of identifying a relationship that exists or the probability of rejecting the null hypothesis when it is false or should be rejected. Let us say you were testing the effectiveness of a new intervention. Without sufficient power to detect a difference between experimental and control group outcomes, there would be no use in conducting the study.

There are basically *four* considerations in determining the size of one's sample:

1. Data analytic procedures that will be used
2. Statistical level of significance (usually chosen at 0.05 or 0.01 level)
3. Statistical power (acceptable range of 0.75 to 0.85)
4. Effect size

Effect size refers to the strength of differences in the sample values that the investigator expects to find. Take, for example, a study to test the effects of a low-impact aerobic exercise program for the elderly. Let us say that based on empirical evidence from literature reviews or data from small scale or pilot studies, the investigator hypothesizes that there would be a large effect of the program on muscle strength and a minimal to moderate effect on cardiovascular fitness. If the difference between values in experimental and control group subjects was large, only a few subjects would be needed in each group to detect such differences. If the effect size is small, a large sample size would be necessary to detect differences.

The number in the sample is also determined by the numbers of units in the sampling frame. Practical considerations such as time and financial support for conducting the research may also provide guidelines on the number of units to be included in the sample.

Summary

We have said that the object of boundary setting in experimental-type designs is twofold: to complete the study in a timely, cost-effective, and manageable manner; and to use the findings about the subset (sample) to inform us about the larger group (population). We presented a number of probability and nonprobability methods that can be used to select a sample from a population. Probability sampling procedures are used to assure that a sample reflects or is representative of the larger population. Nonprobability methods are used when it is not possible to identify a sampling frame or random selection is not feasible or desirable. The sampling process occurs in four steps:

Step 1: Define population
   • Inclusion criteria
   • Exclusion criteria
Step 2: Develop sampling plan
   • Probability or nonprobability
Step 3: Determine sample size
Step 4: Implement sampling procedures and compare critical values of sample to population

Although this four-step process sounds simple, many decisions have to be made along the way. There are many texts available that provide in-depth discussion of each procedural step and the statistical theory of the sampling plans just discussed. Furthermore, the reader should be aware that an investigator may use variations and combinations of these sampling plans, again depending on the nature of the research question and the resources available to the investigator. When probability sampling is used, it is wise to consult a statistical expert to assure that the sampling plan maximizes representation.

In health and human service research, it is frequently not possible to use probability sampling techniques. The state of the field is such that most researchers use nonprobability methods to obtain their sample. However, even with limited external validity, studies that are well planned and in which conclusions are consistent with the level of knowledge that the sampling can yield are extremely valuable.

BOUNDARY SETTING FOR NATURALISTIC DESIGNS

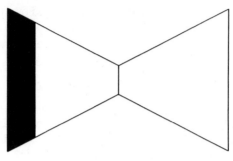

DOMAIN SETTING

Let us now look at how a researcher sets boundaries when conducting a study using naturalistic designs. In naturalistic research, setting boundaries is a dynamic, inductive process. Because the basic purpose of naturalistic research is exploration, understanding, and description, the researcher often does not know the boundaries of the research population before undertaking the study. Indeed, the very point of the study might be to develop the characteristics that bound or define a population. Thus, naturalistic researchers often find that the boundaries themselves emerge from the process of data collection and ongoing analysis.

Let us once again consider the famous work *Tally's Corner* written in 1967 by Elliot Liebow[7] as an example. In his work, Liebow[7] conducted an ethnography to "gain . . . a clear, first-hand picture of lower class Negro men—especially street corner Negroes" [p. 10]. Thus, the boundaries of the research were not clearly delineated by population characteristics, because the researcher did not know them. Rather, the research was initially bounded by the geographic location of a neighborhood in which men lived, worked, and/or "hung out." As the study proceeded, Liebow[7] noted that his data collection took him beyond the initial geographic boundaries of the study to "courtrooms, jails, hospitals, dance halls, beaches and private houses" [p 12]. Liebow initially set boundaries for entry into the study and then expanded and modified these boundaries as the study proceeded. This process is characteristic of boundary setting in naturalistic inquiry.

The researcher has to start somewhere and must make decisions as to what to observe, who to talk to, and how to proceed in the field. How are such decisions made? First, the type of boundary selection is determined by the research problem and the nature of the field study that the researcher develops. In *Tally's Corner*,[7] for example, the geographic location initially formed the boundary because the query sought to understand the lives and experiences of black men who frequented urban street corners.

In other naturalistic studies, cultures may form the initial boundary of the study. A culture may be loosely defined as the customs and tacit knowledge held by a group of individuals who belong to a group, or the culture may be identified by the location in which it exists. For example, the American culture is located in the United States. However, this definition is only a preliminary notion of the boundaries of the inquiry. As the researcher collects information about the culture, he or she refines the boundaries of the study to obtain a fuller understanding of the group.

The process of boundary setting in naturalistic research begins with the investigator determining an entry point into the inquiry. The actual point of entry into the study is often referred to as "gaining access."[8] Often an investigator will seek introduction to a group through one of its members. This member acts as a facilitator or bridge between the life of the group and that of the investigator. From that entry point, which could be, but not limited to, either establishing rapport with other group members, examining a concept, or observing a location in which a culture lives and performs, the researcher collects information or data to begin to describe the boundaries.

Within this process of discovery and the context of the field, the researcher continually makes boundary decisions as to whom to interview and what to observe. There may be an overwhelming number of observational points and potential individuals to interview. Selection decisions are based on the specific questions the researcher poses throughout the research process and the practicalities of the field, such as who is actually available or willing to be interviewed or observed.

There is a range of selection strategies available to the investigator who will often use several approaches or a combination of approaches in any one field study, including the use of probability or nonprobability sampling techniques. In this case, boundary setting begins inductively, and then as concepts emerge and theory development proceeds, the researcher assumes a more deductive way of selecting observations, individuals, or artifacts. For example, observational points may be chosen in such a way as to assure *representation* of that which is observed. Let us say you were interested in understanding how patients feel about their hospital experiences. As part of your study, you would purposely select different observational times throughout the day and night to assure representation of time.

Sometimes decisions are initially made as to whom or what not to observe. Throughout fieldwork, the researcher is actively determining which sources will help the most in understanding the particular question to derive a complete perspective of the culture or practices of the group.

In other types of field studies, boundaries lie within the realm of experiences as described by the individual. For example, in phenomenologic studies the goal is to understand the experiences of a small set of individuals from their own perspective. A phenomenologic researcher who wishes to describe the experience of a parent who has a child with a terminal illness may choose to interview and observe that parent. The boundary of the study, and thus the understanding derived from it, applies only to that individual's lived experience. But once again, the researcher must begin somewhere and make a selection decision as to which individual or individuals to interview who possess the specific characteristic of having a child with the illness. Here the selection process may be *purposeful* in that the researcher makes the judgment as to which parent may be a reasonable informant. Or the researcher may make a selection based on *convenience*, such as any individual who is experiencing life with a child with a terminal illness at that moment and is accessible to the investigator. Or the researcher may interview a parent who has never experienced a child with terminal illness as a way of contrasting and confirming and highlighting the experiences of parents who do.

Thus, once in the field the investigator uses other techniques and approaches to establish the boundaries of the study. For example, to understand certain practices, an ethnographer may seek out a key actor or informant to represent the group. This individual is selected because of either his or her strategic position in the group or because he or she may be more expressive or able to articulate that which the researcher is interested in knowing. Throughout Liebow's[7] study, four key informants continued to provide substantial insights used as the basis of analysis, although many other men participated in the study. The selection of such informants was purposive and based on the judgment of the investigator.

The investigator often starts broadly, sampling, so to speak, all events and individuals within view. Then as fieldwork and interviewing proceed, questions and observational points become more focused and narrow. The investigator may use a snowball or chain approach to identify informants, may look for extreme or deviant cases, may purposely sample diverse individuals or situations to increase variation, and so forth.

It is also possible to set conceptual boundaries. For example, the study by DePoy[9] using a Delphi technique sought to define the concept of "mastery" in occupational therapy practice. The investigator wanted to know what is meant by being "good" at clinical practice. In the first part of a two-part study, DePoy used a nonprobability purposive sampling process to select a group of "experts" in occupational therapy. She then interviewed these experts to obtain an understanding of their meanings of the concept of mastery. However, even though a sample of experts provided data, in this study, the boundary was the concept of mastery, not the individuals who contributed to its definition. In part 2 of the study, DePoy expanded the boundary to the practice arena and examined how mastery was demonstrated in clinical practice. Three informants were selected based on their years of clinical experience and their reputations as excellent clinicians and formed the boundary of the study. The concept of mastery derived from part 1 of the study was then examined in the practices of these three informants through extensive unstructured interview with each informant. DePoy made no claim of external validity but suggested that the model of mastery that was developed in the domain of study be tested for its relevance to other domains.

Let us look at another example of naturalistic research to further understand the boundary setting process. In his study of the culture of a nursing home, Savishinsky[10] originally was interested in examining the behavioral responses of nursing home residents to pet therapy. As he[10] states:

> The approach that I took began with two concepts at the very heart of pet therapy-companionship and domesticity. In the broadest terms, I came to realize that the study had to be a cultural and not just a behavioral one. . . . It had to look at meanings and not simply actions [p 16].

As you can see, the initial boundaries of the study, that of observing actions during pet therapy, became modified and expanded as the study proceeded to the nursing home culture as a whole, including not only behavioral patterns but meanings embedded within that culture.

The process of selecting and adding pieces of information through interview and observation and analyzing them for further discovery and description of the boundaries of the study is called *domain analysis*. Understanding the domain is critical in naturalistic inquiry. The intent is to understand, in-depth, specific phenomena that occur within the domain of the study. Thus, the domain must be clearly understood for findings or interpretations to be meaningful. Although in naturalistic inquiry no claim is made about the representativeness of a domain, principles are suggested that may be relevant to other settings. Thus, naturalistic boundary setting is unlike boundary setting in experimental-type design in which the purpose is to maximize representation and external validity. In naturalistic inquiry, selection processes are used to obtain a "thick description"[11] of a domain.

As indicated by Liebow,[7] the findings in *Tally's Corner* may also explain the phenomenon of the black street corner culture in other settings. However, Liebow himself makes no claim to that effect and suggests that further study is warranted to determine universal commonalities in similar domains. The process of obtaining a "thick description"[11] of a domain allows the investigator to examine the transferability of principles in one domain to another.

Summary

Boundary setting in naturalistic research is an ongoing and active process that emerges inductively from data collection itself. Research is undertaken to not only address a research problem but also to promote further understanding and descriptions of the boundaries of the research. The researcher is constantly making decisions and judgments about who should be interviewed and when and what to observe. The investigator uses a range of selection strategies in a study. The selection strategy used to choose an individual to interview, an event to observe, or artifact to review is based on the specific focus of the investigator. These strategies may include both probability and nonprobability techniques or many others that have been developed specifically for naturalistic inquiry (such as selection of "typical cases" and deviant cases). Although initially a boundary may be defined globally, the investigator delimits the nature of the group, phenomenon, or concept under investigation through domain analysis.

SUBJECTS, RESPONDENTS, INFORMANTS, OR PARTICIPANTS?

Are they subjects, respondents, informants, or participants?[12] These four terms all refer to individuals who agree to become part of a research study. Each term reflects a different way that an individual participates in a research study and the type of relationship formed between the individual and investigator. In experimental-type research, individuals are usually referred to as subjects, a term that denotes their passive role. In survey research, individuals are often referred to as respondents, because they are asked to respond to very specific questions. In naturalistic inquiry, individuals are usually referred to as informants, a term that reflects the active role of informing the investigator as to the context and its cultural rules. Participants can refer to those individuals who enter a "collaborative" relationship with the investigator and contribute to decision making regarding the research process, as well as informing the investigator about themselves. This term is often used in endogenous and participatory action research. Although there are no hard and fast rules in regard to the use of these terms, you should select the one that reflects your preferred way of knowing and the role that individuals have in your study design.

SUMMARY

On the experimental-type design continuum, boundary setting involves a deductive process in which the investigator determines population characteristics and then selects a sample to study. The selection process involves either a plan developed before conducting the research using probability or nonprobability sampling methods. On the naturalistic continuum, boundary setting occurs inductively in which the "domain" is revealed through ongoing data collection and analysis. A range of selection strategies may be used throughout the course of fieldwork. It is important to understand that there is nothing inherently superior about any one of the boundary-setting methods we have discussed in this chapter. The strengths and limitations of a boundary-setting technique depend on its "appropriateness" and "adequacy" in how well it fits the context of the particular research problem.

Appropriateness is defined as the extent to which the method of boundary setting fits the overall purpose of the study, as determined by the research problem and the structure of the research design. For example, it would be inappropriate to use a random sampling technique when one's purpose is to understand how individuals interpret their experience of illness. In this case, the *purposeful* selection of individuals who can articulate their feelings may provide greater insights than the inclusion of a sample with predetermined character- istics. Purposive selection to bound the study facilitates understanding, which is the underlying goal of the study.

By *adequacy* we mean the extent to which the boundary setting yields sufficient data to answer the research problem. In experimental-type designs, adequacy is determined by sample size. In naturalistic designs, adequacy is determined by the quality and completeness of the information and under- standing obtained within the selected domain. *Saturation* in the field obtaining repetitive information and obtaining a thick description of repetitive information are used as cues that the boundary has been obtained and the investigator has achieved a sense of completeness. Thus, appropriateness and adequacy are two criteria that can be applied to boundary-setting decisions.

EXERCISES

1. To understand boundary setting in naturalistic inquiry, select a public place such as a shopping mall or a restaurant. Spend at least 1 hour observing and try to determine patterns of human behavior in the location. As you observe, record how and why you selected to focus on specific elements of the location. Reflect on the range of strategies you used and what each provided in terms of understanding the setting.

2. Develop a sampling plan for a study on lifestyles of adults who are hospitalized for chronic heart disease. First, identify population parameters and then suggest how best to obtain a representative sample of that population.

3. Discuss how you might obtain a random sample of children, given the regulation that participation in research must be consented to by a child's legal guardian.

REFERENCES

1. Krefting L: Reintegration into the community after head injury: the results of an ethnographic study, *Occup Ther J Res* 9:67–83, 1989.

2. American Psychiatric Association: *Diagnostic and statistical manual (DSMIII-R)*, ed 3, revised, Washington, DC, 1988, American Psychiatric Press.
3. Santiago JM, Bachrach LL, Berren MR, et al: Defining the homeless mentally ill: a methodological note, *Hosp Community Psychiatry* 39:1100–1102, 1989.
4. Bailey D: Reasons for attrition from occupational therapy, *Am J Occup Ther* 44:23–29, 1990.
5. Suffet F, Liftshitz M: Women addicts and the threat of AIDS, *Qualitative Health Res* 1:51–79, 1991.
6. Estes J, Deyer C, Hansen R, et al: Influence of occupational therapy curricula on students' attitudes toward persons with disabilities, *Am J Occup Ther* 45:156–165, 1991.
7. Liebow E: *Tally's corner*, Boston, 1967, Little, Brown.
8. Spradley J: *Participant observation*, New York, 1980, Holt, Rinehart & Winston.
9. DePoy E: Mastery in clinical occupational therapy, *Am J Occup Ther* 44:415–420, 1990.
10. Savishinsky JS: *The ends of time*, New York, 1991, Bergen & Garvey.
11. Geertz C: *The interpretation of cultures: selected essays*, New York, 1973, Basic Books.
12. Subjects, respondents, informants and participants [editorial], *Qualitative Health Res* 1:403–406, 1991.

13

Collecting Information

- **Principles of Collection Process**
- **Summary**

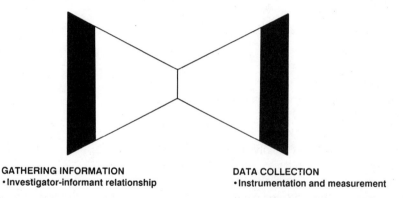

GATHERING INFORMATION
- Investigator-informant relationship

DATA COLLECTION
- Instrumentation and measurement

Finally we are ready for action. This chapter discusses the range of strategies and procedures researchers implement to obtain information to answer the question, or query, they have posed. We label this phase of the research process as *data collection* to reflect the actions engaged in along the experimental-type continuum and *gathering* or *collecting information* to reflect naturalistic design activity. We will discuss a variety of collection procedures. Some are used by designs along both continua, and some are specific to a paradigmatic framework and design category.

In the experimental-type continuum, the data collection phase of research represents the crossroads between question formation and the development and implementation of design and analysis. Strategies for data collection involve actions that reflect the research problem and design. In turn, the type of data collected and the way data are obtained shape the type and nature of the analytic process and the understandings and knowledge that can emerge. In naturalistic inquiry, gathering information stems from the initial query but is part of a

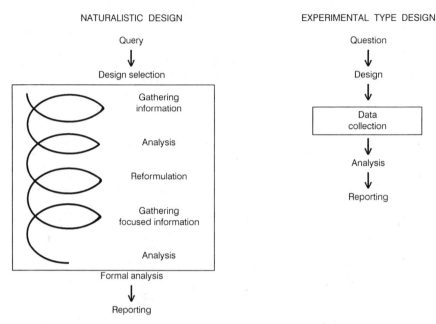

NATURALISTIC DESIGN   EXPERIMENTAL TYPE DESIGN

FIG 13–1.
Naturalistic design and experimental-type design.

dynamic process that involves ongoing analysis and the reformulation of the initial query.

Figure 13–1 displays the nature of collecting information in each design continuum. In this chapter we examine the general principles and strategies of data and information collection that inform both research continua. Then in Chapter 14 we discuss the specific approaches used by researchers along the experimental-type continuum, and in Chapter 15 we focus on the action processes in naturalistic inquiry.

PRINCIPLES OF COLLECTION PROCESS

Three basic principles organize the collection process across the two research design continua. First, the aim of collecting information or data, regardless of how it is done, is to obtain information that is both relevant and sufficient to answer a research question or query. Second, as summarized in Table 13–1, the choice of a data collection or gathering information strategy is based on four major factors: the researcher's paradigmatic framework, the nature of the research problem, the type of design, and the practical limitations

TABLE 13–1.

Factors Influencing Choice of Collection Strategy

Factor	Specific Issue
Paradigmatic framework	Relationship of investigator to that which is investigated
Research problem	Question or query
Design type	Placement of design along a continuum
Practical limitations	Time
	Money
	Access to research population

or resources available to the investigator. Third, although a collection strategy reflects one's basic paradigmatic framework, a specific procedure, such as observation or interview, may be shared by researchers from either a naturalistic or positivist paradigm. Many researchers find it useful to collect data and information with more than one procedure or technique to answer a question or query more fully. The use of multiple collection techniques is called *triangulation of method*[1] and is discussed in subsequent chapters.

In general, data and information collection is comprised of one or more of the following strategies: (1) watching and listening, (2) asking, and (3) obtaining and examining materials. As summarized in Table 13–2, each one of these strategies can be structured, semistructured, or open ended, depending on the nature of the inquiry. Also, each strategy can be used in combination with the other. Let us examine each strategy.

Watching and Listening

Watching and listening in research is the process of *systematic observation,* which includes three activities: watching, listening, and recording data. The way in which one watches and listens may range in approach from structured to unstructured, participatory to nonparticipatory, focused to broad, and time limited to fully immersed.[2] Along the experimental-type continuum, observation is

TABLE 13–2.

Summary of Collection Methods

Method	Structured	Unstructured
Watching and listening	Checklists	Participatory
	Rating scales	Nonparticipatory
Asking	Closed-ended	Open-ended
		Guided questioning
Artifact	Coding schemes	Open-ended observation

usually time limited and structured. Criteria to watch and listen are selected a priori to gathering data, and the data are recorded along a structured measurement system. Checklists may be used that indicate the presence or absence of a particular behavior or object under study. Phenomena other than those specified before entering the field are ignored or remain outside the investigator's interest. Rating scales may also be used to record observations in which the investigator rates the observed phenomenon on a scale ranging on a predetermined point system. We discuss rating scales in more detail in Chapter 14.

In naturalistic designs, watching and listening take on an inductive, participatory quality. The investigator broadly defines the field of observation (such as a nursing home), and then moves to a more focused observational approach as the process of collecting and analyzing information from previous observations unfolds.

Let us look at the concept of intelligence to illustrate different observational approaches. In experimental-type studies, intelligence testing, in large part, relies on the use of a structured instrument that involves paper and pencil tests and a tester observing a subject's performance in a laboratory-type setting. Measurements are obtained on dimensions that have been predefined as comprising the construct of "intelligence." However, in a naturalistic approach, rather than use predefined criteria, the investigator might watch and listen to persons in their natural environments to reveal the meaning of intelligence in a particular culture. Each approach has its value in addressing specific types of research problems. Structured observation of intelligence would be appropriate for the investigator who wishes to compare populations or individuals, measure individual or group progress and development, or describe population parameters on a standard indicator of intelligence. On the other hand, naturalistic observation of intelligence would be useful to the researcher who is attempting to develop new understandings of the construct in populations such as certain minority groups in the United States, the elderly or non-Western cultures in which the standard intelligence tests do not seem to apply or fit.

Asking

Health and human service professionals routinely ask questions to obtain information. In research, asking is a systematic and purposeful aspect of a data collection plan. As in observation, asking questions can vary in structure and content from unstructured open-ended to closed-ended or fixed response questions that use a predetermined response set. An example of an open-ended question would be, "How have you have been feeling this past week?" In contrast, a structured or fixed response approach would be, "In this past week, how would you rate your health: excellent, good, adequate, fair, or poor?" Of no surprise, naturalistic research relies more heavily on open-ended types of asking techniques, whereas experimental-type designs tend to use structured fixed response questions. Focused, structured asking is used to obtain data on a specified phenomenon, whereas open-ended asking is used when the research purpose is discovery and exploration.

When one is asking, questions can be posed either through *interview* or *questionnaires.*

Interviews

Interviews are conducted through verbal communication and may occur face to face or by telephone and be either structured or unstructured. Interviews are usually conducted with one individual. However, sometimes group interviews are appropriate such as those conducted with couples, families, and work groups. Group interviews of five or more individuals involve a focus group methodology in which the interaction of individuals is key to the information the investigator wishes to obtain.[3] Audiotape of the group interview is transcribed, and the narrative, along with investigator notes, forms the information base, which is then analyzed (see Chapter 11 for a discussion of this design type).

Structured interviews involve a written document in which maximum researcher control is imposed in the content and sequencing of questions. That is, each question and its response alternatives are developed and placed in an order before an interview is conducted. Interviewers are instructed to ask subjects each question precisely as it is written. Most questions are closed ended, or "fixed," in that subjects select one response from a predetermined set of answers. Closed-ended questions vary with regard to the type of response alternatives. They can be either:

1. Dichotomous (yes/no)
2. Multiple choice
3. Rank order questions (Guttman scale)
4. Five or more responses (Likert scale)

Each response set forms a different type of scale, and these are discussed in Chapter 14.

In a structured interview, sometimes the investigator may use a few open-ended questions. Responses are examined and coded and in effect are changed into a closed-ended response set on a posthoc basis. That is, the investigator develops a numeric coding scheme based on the range of responses obtained and then assigns a code to each subject. The numeric response is then analyzed.

Unstructured interviews are used primarily in naturalistic research and exploratory-type studies in experimental-type designs. The researcher initially presents the topic area of the interview to a respondent and then uses probing questions to obtain information. The interview may begin with an explanation of the study purpose and a broad statement or question such as, "Could you please describe your daily experiences as a caregiver?" Other probing questions emerge as a consequence of the information provided from this initial query. A *probe* is a statement that is neutral or does not bias the subject to respond in any particular way. Probes are used to encourage the respondent to provide more information or elaborations. A probe such as "tell me more about it" or sometimes just repeating a question serves as a way of encouraging a respondent to discuss an issue further. Unstructured interviews are sometimes used by quantitative researchers in a pilot study to uncover domains and response codes for inclusion in a more structured interview and question format.

Questionnaires

Questionnaires are written instruments and may be administered face to face, by proxy, or through the mail. Like interviews, questionnaires vary in their level of structure of questions used. The process of developing questionnaires is examined in Chapter 14 as part of the discussion on measurement in experimental-type research.

Each way of asking, structured or unstructured, has strengths and limitations that the researcher needs to understand and weigh to determine the most appropriate approach to take. Strengths and limitations must be evaluated in terms of the researcher's purpose. Let us look at a study conducted by Atchison et al.[4] to illustrate this point. In this study, the investigators were interested in understanding the attitudes, knowledge and fears of occupational therapists regarding human immunodeficiency virus (HIV) and acquired immunodeficiency syndrome (AIDS). They chose to ask about these fears by using a structured questionnaire format that had been previously used to ask nursing students the same questions. Respondents had to rate their agreement with items on a 5-point Likert scale (strongly agree, agree, uncertain, disagree, strongly disagree). This type of asking limited the response of subjects to only these five options. Any other comments or explanations were considered extraneous and were not recorded. The advantages of this type of asking in this study are:

1. Statistical analysis can be used to obtain an understanding of group responses.
2. Respondents might be more likely to respond honestly on paper knowing that their confidentiality will be upheld.
3. Large cohort of persons (in this case a sample size of 79) participate in this type of asking in a relatively short period of time.
4. Responses from this group can be compared with another group because the test can be repeated in the same way.
5. Researcher obtains responses to only those questions of interest.

Limitations of this approach are:

1. Researcher may not capture the issues that are relevant to occupational therapists.
2. Researcher remains unsure if respondents answer in a socially desirable way rather than indicating how they actually feel.

Let us consider a different way of asking respondents about HIV and AIDS. Using an open-ended, face-to-face interview approach, a researcher may obtain more in-depth and relevant information that may not have been included in a structured questionnaire. Further, the complexity of responses to HIV and AIDS would be retained and become the focus for analysis. The advantages of using open-ended asking include:

1. Interviewer can develop rapport that may be important to explore a highly sensitive issue.
2. Nonverbal information that is communicated and may be important can be captured and analyzed.
3. Researcher can determine which issues are most important or relevant to the respondent.

The limitations of using an open-ended questioning approach in the study might include:

1. Respondents who do not want to discuss such sensitive issues directly may resist.
2. Additional time would be required to obtain interviews with a large sample and to analyze the information.

As you can see, both types of asking have their merits and limitations, depending on the research purpose, phenomena to be studied, and the study population.

Obtaining and Examining Materials

There are numerous reasons why researchers use existing data. First, this type of information may be accessible where direct observation and interview data are not. Second, the use of existing materials eliminates the attention factor, or the Hawthorne effect.[5] This effect refers to a change in a respondent's answer as a consequence of participating in the research process itself. Third, use of existing data allows the researcher to view phenomena in the past or over time, which might not be possible with primary data collected at one point in time. Finally, existing materials are valuable in obtaining data in sensitive inquiries, where informants may not want to share accountings or may not be able to do so.

In experimental-type design, securing and examining materials are structured by criteria before the research field is entered. In naturalistic inquiry, the selection of materials to observe or examine emerges as a consequence of the investigative process. There are three distinct approaches to obtaining and examining existing materials: (1) unobtrusive methodology, (2) secondary data analysis, and (3) artifact review.

Unobtrusive Methodology

Unobtrusive, or nonreactive, methodology[6] involves the observation and examination of documents and objects that bear on the phenomenon of interest. For example, to estimate the alcohol consumption of a neighborhood, the investigator might look through trash cans and count the empty alcohol containers. This approach can be used to confirm information obtained from another source. So in this example, the researcher might have looked through trash cans as a way of confirming information obtained through a face-to-face interview on alcohol consumption. Examples of written materials that investigators may use for analysis include:

- Diaries/journals
- Medical records
- Clinical notes
- Historical documents
- Minutes from meetings
- Letters

Secondary Data Analysis

In secondary data analysis, the researcher "reanalyzes" one or more existing data sets.[7] For example, a health care researcher might look at patient medical records to obtain data or might combine the data of two studies for subsequent analysis. A social work researcher may examine process recordings as a basis for understanding therapeutic interaction in family therapy. The purpose of a secondary analysis is to ask different questions of the data from that analyzed in the original work. There are several large national health and social service data sets that researchers use for secondary data analysis, such as the census tracts, National Health Care Data Set, Medicare and Medicaid data, and court report recordings that are important sources for secondary analysis in health and human service research.

Artifact Review

Artifact review is a technique used primarily to ascertain the meaning of objects in research contexts. For example, archaeologists examine ruins to learn about ancient cultures. Artifact review in health and human service research might include the examination of personal objects in a patient's hospital room or client's home to determine the interests, socioeconomic status, and so forth, as a basis for intervention planning.

SUMMARY

We have described three basic principles that guide the action process of obtaining information: (1) Collect relevant and sufficient information for the research question or query. (2) Choose an information gathering or data collecting strategy that is consistent with the research question, epistemologic foundation, design, and practical constraints of the research effort. (3) Recognize that methods can be shared across continua. We then discussed the three categories of collecting information in all designs: watching and listening, asking questions, and examining materials. You are now ready to move on to Chapters 14 and 15, which examine specific data collection strategies in experimental-type design and gathering information in naturalistic design.

REFERENCES

1. Morse JM: *Qualitative nursing research,* Rockville, Md, 1989, Aspen.
2. Patton M: *Qualitative evaluation and research methods,* ed 2, Newbury Park, Calif, 1990, Sage.
3. Kreuger R: *Focus groups: a practical guide for applied research,* Newbury Park, Calif, 1988, Sage.
4. Atchison DJ, Beard BJ, Lester LB: Occupational therapy personnel and AIDS: attitudes, knowledge and fears, *Am J Occup Ther* 44:212–217, 1990.
5. Roethlisberger FJ, Dickson WJ: *Management and the worker,* Cambridge, Mass, 1947, Harvard University Press.
6. Webb EJ, Campbell DT, Schwartz RD, et al: *Nonreactive measures in the social sciences,* Boston, 1981, Houghton Mifflin.
7. Stewart DW: *Secondary research: information sources and methods,* Newsbury, Calif, 1984, Sage.

14 | Data Collection for Experimental-Type Research: Concern With Measurement

- **Measurement Process**
- **Confidence in Instruments: Reliability and Validity**
- **Instrument Construction**
- **Administering the Instrument**
- **Summary**
- **Exercises**

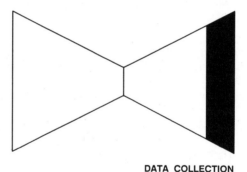

DATA COLLECTION

In experimental-type design, the purpose of data collection is to find out about an "objective reality". As we indicated in previous chapters, the investigator is considered to be separate and removed from that which is to be known. There is strict adherence to a data collection protocol. A protocol refers to a series of procedures and techniques designed to remove the influence of the investigator from the data collection process and assure a nonbiased and uniform approach to obtain information. This information or data are further objectified by the assignment of numeric values. These numeric values are then submitted to statistical procedures to test relationships, hypotheses, and population descriptors. These descriptions are viewed as representing a reality or "objective truth."

Therefore, in experimental-type research, the process of quantifying information, or "measurement," is a primary concern. The investigator must develop instruments that are reliable and valid or have a degree of correspondence to an objective world or truth.

In this chapter we discuss the key issues of data collection for experimental-type research. These are the measurement process and instrumentation or issues of reliability and validity.

MEASUREMENT PROCESS

Measurement, defined as the translation of observations into numbers, is a vital action process in data collection in experimental-type research. This action process links the researcher's abstractions or theoretical concepts to concrete variables that can be empirically or "objectively" examined. In experimental-type research, concepts must be operationalized or put into a format, such as a questionnaire or interview schedule, that will permit some type of structured, controlled observation or measurement. Thus, measurement involves both conceptual and operational or empirical considerations, as displayed in Figure 14–1.

The conceptual or theoretical consideration first involves identifying a concept and then defining it. It involves asking, "How shall I conceptualize the phenomenon under study such as depression, anxiety, sense of personal mastery, or adaptation?" Or, "How shall I define these concepts in words?"

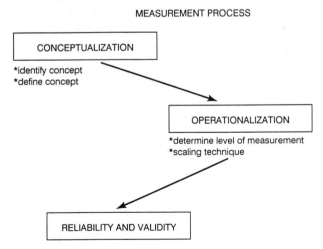

MEASUREMENT PROCESS

CONCEPTUALIZATION

*identify concept
*define concept

OPERATIONALIZATION

*determine level of measurement
*scaling technique

RELIABILITY AND VALIDITY

FIG 14–1.
Measurement process.

The operational, practical or empirical consideration of measurement involves asking, "What kind of an indicator shall I develop or use as a gauge of this concept?" Or, "How shall I classify and or quantify what I observe?" An indicator or measure is an empirical representation of an underlying concept. Because, by definition, a concept is always unobservable, the strength of the relationship between indicator and concept is critical. The more consistent or reliable and the more accurate or valid the relationship, the more desirable the measuring procedure or instrument. Thus, reliability and validity are two fundamental properties of indicators. Researchers engage in major efforts to evaluate the strength of reliability and validity to determine the desirability and value of a measure.

Thus, the action process of information gathering in experimental-type research begins with the identification of concepts and then proceeds to derive both conceptual and operational definitions of those concepts. Based on these definitions, specific indicators (scales, questionnaires, rating forms) are specified or developed and determine the information that will be collected.

Development of Indicators: Levels of Measurement

The first step in the development of an indicator involves specifying how a variable will be operationalized. This process involves determining the numeric level at which a phenomenon can be measured (nominal, ordinal, interval, or ratio). By level of measurement we mean the properties and meaning of the numbers assigned to an observation and further the boundaries regarding the type of mathematical manipulations that can be performed with those numbers. Determining how a variable is measured directly shapes the type of statistical analysis that can be performed and is an important part of the measurement process.

Every variable must have two qualities. First, the attributes of a variable should be *exhaustive*. That is, the variable should be able to classify every observation in terms of one of the defined attributes of that variable. A simple example is the concept of gender. The categories of male and female represent the full range of attributes for this concept. The second quality of a variable is that the attributes or categories must be *mutually exclusive*. For example, one can be only male or female.[1]

Variables operationalize the concepts we wish to study, and they can be either discrete or continuous. *Discrete variables* are those with a finite number of distinct values.[2] Gender, with only two possible values, represents a discrete variable in that one is either male or female. There is no in-between category. *Continuous variables*,[2] on the other hand, can take on an infinite number of values. Age and height for example, represent continuous variables in that they can be measured along a numeric continuum of years, months, days, or inches and feet, respectively.

It is possible to measure a continuous variable using discrete categories, such as the classification of the variable "age" as young, middle-aged, and old or the variable "height" as tall, medium, or short. Discrete and continuous structures represent the natural characteristics of a variable. These characteristics have a direct bearing on the level of measurement or the rules of measure that can be applied. Let us examine each level of measurement.

Nominal

The simplest level of measurement is nominal, which involves classifying observations into mutually exclusive categories. The word nominal means name, and therefore, at this level of measurement numbers are used, in essence, to name attributes of a variable. For example, your telephone number, the number on a football jersey, or the number on your social security card is used to identify you as the attribute of the variables "persons with a telephone," "football player," and "taxpayer," respectively. All are examples of nominal numbers.

Variables at the nominal level are discrete. For example, we might classify individuals according to their political or religious affiliation, gender, or ethnicity. In each of these examples, membership in one category excludes membership in another. Examples of mutually exclusive categories include:

- Yes or no
- Male or female
- Democrat, Republican, or independent
- White, black, Hispanic, or Asian

For data analysis, the researcher assigns a numeric value to each nominal category. For example, "male" may be assigned a value of 1, whereas "female" may be assigned a value of 2. The assignment of numbers is purely arbitrary, and no mathematical functions or assumptions of magnitude or ranking are implied or can be performed.

Ordinal

The second level of measurement is ordinal, which involves the ranking of phenomena. The word "ordinal" means order and thus can be remembered as the numeric value that assigns an order to a set of observations. Variables that are discrete and can be conceptualized as having an inherent order at the theoretical level can be operationalized in a rank order format. Variables operationalized at this level have the same properties of nominal categories in that each category is mutually exclusive of the other. In addition, ordinal measures have a fixed order or ranking, so that one can say that one category is higher or lower than another category. Take, for example, the ranking of income into categories such as:

1 = Poor
2 = lower income
3 = middle income
4 = upper income

We can say that middle income is ranked higher than lower income, but we can say nothing about the extent to which they differ. The assignment of a numeric value is symbolic and arbitrary as in the case of nominal variables because the distance or spacing between each category is not numerically equivalent. However, the numbers do imply magnitude, that is, one is greater than the other. Because there are no equal intervals between ordinal numbers, mathematical functions such as adding, subtracting, dividing, and multiplying cannot be performed. One can merely state that one category is higher or lower, stronger or weaker, or greater or lesser.

Interval

The next level of measurement shares the characteristics of ordinal and nominal measures but also has the characteristic of equal spacing between categories. Interval measures are continuous variables in which the zero point is arbitrary. Although of a higher order than ordinal and nominal measure, the absence of a true zero point does not allow statements concerning ratios to be made. However, the equidistance between points allows the researcher to say that the difference between scores of 50 and 70 is equivalent to the difference in scores of 20 and 40. Examples of a true interval level of measurement includes Fahrenheit and Celsius temperature scales or IQ scales. In each of these cases, there is no absolute zero, but there is equal distancing between mutually exclusive categories. In the social and behavioral sciences, there is considerable debate as to whether behavioral scales represent interval levels of measurement. Such scales typically have a Likert-type response format in which a study participant responds to one of five or seven categories such as strongly agree, agree, uncertain, disagree, or strongly disagree. Those who accept this type of scaling as an interval measure argue that the distance between strongly agree and agree is equivalent to the distance between disagree and strongly disagree. Others argue that there is no empirical justification for making this assumption and that the data generated should be considered as ordinal in nature. In the actual practice of research, many investigators assume such scales are at the interval level to use more sophisticated and powerful statistical procedures that are possible only with interval and ratio data. You should be aware that this debate continues to be a controversial matter among experimental-type design researchers and becomes extremely important when data analysis techniques are selected.

Ratio

Ratio measures represent the highest level of measurement. Such measures have all the characteristics of the previous levels and, in addition, have an absolute zero point. Income is an example of a ratio measure. In this case, instead of classifying income into ordinal categories, as in the previous example, it is described in terms of dollars. Income is a ratio measurement because it is possible for someone to have an income of 0, and we can say that an income of $40,000 is twice as high as an income of $20,000.

Table 14–1 summarizes the characteristics of each level of measurement.

Experimental-type researchers usually strive to measure a variable at its highest possible level. However, the level of measure is also a reflection of the researcher's concepts. If, for example, political affiliation, gender, or ethnicity are the concepts of interest, nominal measurement may be the most appropriate level, because magnitude does not apply to those concepts.

Types of Measures

Now that we have discussed the properties of numbers, let us examine the different methods by which data are collected and transformed into numbers. We begin with a discussion of specific types of measures that are frequently used in experimental-type research.

Scales

Scales are tools for the quantitative measurement of the degree to which individuals possess a specific attribute or trait. Here are only a few examples of scales that are frequently used by health and human service professionals in research:

- Dyadic Adjustment Scale (a measure of marital adjustment)[3]
- Self-Rating Anxiety Scale (a measure of clinical anxiety)[4]

TABLE 14–1.

Characteristics of Levels of Measurement

	Nominal	Ordinal	Interval	Ratio
Mutually exclusive categories	X	X	X	X
Fixed ordering		X	X	X
Equal spacing			X	X
Absolute zero				X

- Family Adaptability Cohesion Scale[5]
- State-Trait Anxiety Scale[6]
- Session Evaluation Questionnaire[7]
- CES-D Depression Scale[8]

The three basic approaches to scaling are Likert, Guttman, and semantic differential.[9]

Likert Scale.—In this approach to scales, the researcher develops a series of items (usually between 10 and 20) worded favorably and unfavorably regarding the underlying construct that is to be assessed. Respondents then indicate a level of agreement or disagreement with each statement by selecting one of several, usually five to seven, response alternatives. Researchers may then combine responses to the questions to obtain a summated score, may examine each item separately, or summate scores on specific groups of items to create subindexes. In developing a Likert-type format the researcher must decide how many response categories to allow and whether the categories should be even or odd in number. Even choices force a positive or negative response, whereas odd numbers allow the respondent to select a neutral or "middle" response. The Dyadic Adjustment Scale[3] is an example of a Likert-type scale with several different even-numbered response formats. The first set of items (22 separate items) have 6 response categories that range from always agree, agree a lot, agree a little, disagree a little, disagree a lot, to always agree. The instructions read:

> Most persons have disagreements in their relationships. Please indicate below the approximate extent of agreement or disagreement between you and your partner for each item on the following list.

The FACES-III[5] (Family Adaptability and Cohesion Scale), on the other hand, is an example of a scale measuring family function that allows a neutral response by structuring responses into five ordinal categories: almost never, once in a while, sometimes, frequently, and almost always. The "sometimes" category is the middle response and receives a score of 3.

Likert-type scaling techniques are used frequently by social science researchers. They provide a closed-ended set of responses while still giving the respondent a reasonable range of latitude. Likert-type scales, as we indicated earlier, may be considered ordinal or interval, depending on what is being measured and on the desired data analytic procedures to be selected. A disadvantage of Likert-type scaling is that there is no guarantee that respondents' senses of the magnitude of each response are similar.

Guttman Scale.—This type of scale is referred to as a unidimensional, or cumulative, scale.[1] The researcher develops a small number of items (four or seven) that relate to one concept. The items form a homogeneous or unitary set and are cumulative or graduated in intensity of expression. In other words, the items are arranged hierarchically so that endorsement of one item means an endorsement of those items below it, which are expressed at less intensity. Knowledge of the total score is predictive of the individual's responses to each item. Let us consider a series of questions developed by Stouffer,[10] who was interested in measuring tolerance, about a man who admits he is a communist.

1. Suppose this admitted communist wants to make a speech in your community. Should he be banned from speaking?
2. If some people in your community suggested that a book he wrote favoring government ownership should be taken out of your public library, would you favor removing the book or not?
3. Suppose he is a clerk in a store. Should he be fired or not?

In this series, you can see that if a person says "no" to items 1, 2, and 3, he or she would be most tolerant. On the other hand, a person who says yes to all three items would be the least tolerant. Another person who says "no" to item 2 will also probably have said "no" to item no. 1. Thus, in the Guttman approach to scaling, both persons and items are arranged according to degree of agreement or intensity.

Semantic Differential.—This approach to scaling is usually used for psychologic measures to assess attitudes and beliefs.[11] The researcher develops a series of rating scales in which the respondent is asked to give a judgment about something along an ordered dimension, usually of 7 points. Ratings are bipolar in that they specify two opposite ends of a continuum (good-bad; happy-sad). The researcher then sums the points across the items. The Session Evaluation Questionnaire[7] is an example of a semantic differential scale. The scale measures two constructs related to clients' perceptions of clinical counseling sessions in which they have participated—depth and smooth-ness—and two postsession mood constructs—positivity and arousal. The scale includes 24 items, 12 each for the 2 sections. One item related to the session reads:

> The session was:
> Bad —— Good

One postsession item reads:

> Right now, I feel:
> Happy —— Sad

The respondent is instructed to place an X in the space that most accurately depicts his or her feelings. Semantic differential scaling is most useful when natural phenomena can be categorized into contraries. However, they also limit the range of responses to a linear format.

Summary.—Scaling techniques measure the extent to which respondents possess an attribute or personal characteristic. There are three primary scaling formats used by experimental-type researchers: Likert-type scaling, Guttman scaling, and semantic differential. Each has its merits and disadvantages. The researcher who is developing a scale needs to make some basic formatting decisions about response set structure, whereas the researcher selecting an already existing scale must evaluate the format of the measure for his or her project.

CONFIDENCE IN INSTRUMENTS: RELIABILITY AND VALIDITY

In selecting a scale or other type of measurement, the researcher is concerned about two issues. First, does the instrument measure a variable consistently or reliably? Second, is the instrument an adequate or valid measure of the underlying concept of interest? You might remember that we discussed these two concepts as they related to design in Chapter 7. The two concepts address basically the same issues when referring to instrumentation.

Reliability

Reliability refers to the extent to which you can rely on the results obtained from an instrument. That is, if you were to measure the same variable in the same person in the same situation over and over again, would your result be the same? More formally, reliability refers to the degree of consistency with which an instrument measures an attribute or the ability of an instrument or indicator to produce similar scores on repeated testing occasions that occur under similar conditions. The reliability of an instrument or indicator is important to assess to be assured that variations in the variable under study represent observable variations and not those resulting from the measurement process itself. Refer to Chapter 7 for a refresher on threats to internal design validity, specifically those related to instrumentation. You can see why you would want to be assured that subjects will respond in the same way on repeated testing occasions. If the instrument yielded different scores each time a subject was tested, the scale would not be able to detect the "truth," or objective value, of the phenomenon under investigation.

It is easy to understand the concept of reliability when you consider a physiologic measure such as a scale for weight. If you were to weigh an individual on a scale, you would expect to derive the same weight on repeated measures if that subject stepped on and off the scale several times without doing anything to alter his or her weight in between measurements. Such instruments as a weight scale or blood pressure cuff must be consistent to assess an objective value that approximates a "true" physiologic value for an individual. Let us consider another example. Suppose you are interested in measuring the prelevels and postlevels of depression in a group of inpatients who are in an experimental intervention program. You certainly would want to ensure that a measure was reliable so that any changes observed in depression scores after the intervention would indicate the true alteration in level of depression rather than a variation on a scale.

However, there is always some error in measurement. A scale may not always be completely accurate. Take, for example, a tape measure that is not held precisely at the same place for each measure. This error is random in that it occurs by chance and the nature of the error may vary with each measurement period. Reliability represents an indication of the degree to which such random error exists in a measurement. A formal representation of random error in measurement is expressed by examining the relationship between three components: (1) the true score that is unknown (T); (2) the observed score (O), which is derived from an instrument or another measurement process; and (3) an error score (E). An observed score will be a function of the true score and some error. This tripartite relationship can be expressed as:

$$O = T + E$$

As one can see in this equation, the smaller the error term, the more closely O, or the observed score, will approximate T, the true value. The standardization of study procedures and instruments serves to reduce this random error term. In a questionnaire, for example, a question that is clearly phrased and unambiguous will increase its reliability.

Let us consider an example. Suppose you wanted to know whether your client is feeling depressed. You ask, "How do you feel today?" The client answers, "Fine," and then begins to weep. Obviously the client and you did not have the same question in mind, even though you both heard the same words. If you had asked the client, "Do you feel depressed today?," you would have been more likely to obtain the desired information. Although both questions may mean the same to you, the second question eliminates room for interpretation on the part of the client. The client is then able to understand exactly what you are asking and can give you the answer that you seek.

The same holds for measurement. Reduction of ambiguity decreases the likelihood of misinterpretation and thus of error. Also, the longer the test or more information collected to represent the underlying concept, the more reliable the instrument will become. Let us consider the example of depression once again. If your client answers that he or she is not depressed but breaks down in your clinic and weeps, you might experience some cognitive dissonance. However, if you administered a scale to the client that asked multiple questions that measure depression, you might be able to obtain a more accurate picture of your client's mood.

Because all measurement techniques contain some random error, or error that occurs by chance, reliability can exist only in degrees. The concept of reliability is thus expressed as a ratio between variation surrounding a true score (T) to the variation of an observed score (O):

$$r = \frac{T}{O}$$

Because we can never know the "true" score, reliability represents an approximation to that "objective reality."

Reliability is expressed as a form of a correlation coefficient that ranges from a low of zero to a high of 1.00. The difference between the observed coefficient and 1.00 tells us what percent of the score variance can be attributed to error. For example, if the coefficient is 0.65, then 0.35 reflects the degree of inconsistency of the instrument. On the other hand, 65% of the observed variance is measuring an individual's true, or actual, score.

To assert reliability, experimental-type researchers frequently conduct statistical tests. These tests of reliability focus on three elements: stability, internal consistency, and equivalence. The choice of a test depends on the nature and intended purpose of the instrument.

Tests of Reliability

Stability	Internal Consistency	Equivalence
Test-retest	Split-half Cronbach's alpha Kuder-Richardson formula	Inter-rater Alternate forms

Stability

Stability is concerned with the consistency of repeated measures. Test-retest is a reliability test in which the same test is given twice to the same subject under the same circumstances. It is assumed that the individual should score the same on the test and retest. The degree to which the test and retest scores correlate, or are associated, is an indication of the stability of the instrument and its reliability. This measure of reliability is often used with physical measures and paper and pencil scales. The use of test-retest techniques is based on the assumption that the phenomenon to be measured remains the same at two testing times and that any change is the result of random error. Usually physiologic measures and equipment can be tested for stability in a short time period, whereas paper and pencil tests are retested in a 2-week to 1-month period. This approach to measuring reliability would not be appropriate for instruments in which the phenomenon under study is expected to vary over time. For example, in the measurement of mood or any other transitory characteristic, we would anticipate variations in the true score at each repeated testing occasion. Thus, a test-retest approach would not be an appropriate way of indicating its stability.

Tests of Internal Consistency

These tests are also referred to as "tests of homogeneity" and are used primarily with paper-and-pencil tests. In these tests the issue of consistency is examined internally in relationship to a composite score. In the split-half technique, instrument items are split in half, and a correlational procedure is performed between the two halves. In other words, the scores on one half of the test are compared with scores on the second half. The higher the correlation or relationship between the two scores, the greater the reliability. Tests are split in half by comparing odd-numbered responses to even-numbered responses, by randomly selecting half of the items, or merely by splitting the instrument in half (e.g., questions 1 through 10 are compared with questions 11 through 20). This technique assumes that one would respond consistently throughout a measure and therefore should demonstrate consistent scores between the halves.

Cronback's alpha or Kuder-Richardson formula (K-R 20) are two statistical procedures often used to examine the extent to which all the items in the instrument measure the same construct. The statistic generated is a correlation coefficient between the values of 0 and $+1$, in which a 0.80 correlation is interpreted as adequate reliability. An 80% correlation basically means that the test is consistent 80% of the time and that error may occur 20% of the time. A score as close to $+1$ is most desirable in that such a score indicates minimal error and maximum consistency.

Equivalence

There are two tests of equivalence: inter-rater reliability and alternate forms. *Inter-rater reliability* test involves the comparison of two observers measuring the same event. A test is given to one subject but scored by two or more raters, or the same rater observes and rescores an event on two different occasions. The total number of agreements between raters is compared with the total number of possible agreements and multiplied by 100 to obtain the percent of agreement. Eighty percent or more is considered to be an indication of a reliable instrument. As in the other tests of reliability, a higher score is most desirable.

Consider the following hypothetical data. There are four potential chances for agreement, yet only two actual agreements among the raters. Thus, agreement occurred in the ratio of 2/4 or 1/2. Therefore, the inter-rater reliability in the example is 50%, $1/2 \times 100 = 50$.

Inter-rater Reliability	
Rater 1	Rater 2
30	35
25	25
31	32
33	33

The *alternate, or parallel, forms* test involves the comparison of two versions of the same paper and pencil instrument. In this case two forms of one test that are considered equivalent are administered to one subject on the same testing occasion. The use of this technique depends on the availability of two equivalent versions of a form.

Validity

An indicator or measure not only must be reliable but also must be valid. Validity addresses the critical relationship between a concept and how it is measured. It asks whether what is being measured is a reflection of the underlying concept. Figure 14–2 diagrams three different degrees of validity and highlights how instrumentation (A_1) only approximates the underlying concept (A). The closer an instrument comes to representing the true definition of the concept, the more valid the instrument.

Take, for example, the theoretical concept of intelligence and the traditional indicator of adult intelligence, the Wechler Adult Intelligence Scale.[12] Research-

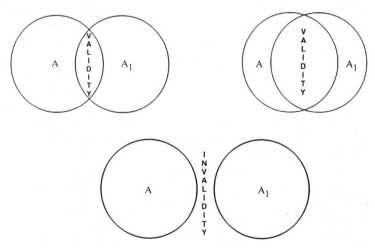

FIG 14–2.
Three degrees of validity.

ers have argued whether this test actually measures what it purports to and whether it is valid for different cultures and populations. Because one can only approximate a perfectly valid instrument, as in reliability, validity is a question of degree and may vary with the intended purpose of the instrument and the specific population for which it is developed or intended. Thus, the Wechler Adult Intelligence Scale may be a valid measure of intelligence for white middle-class Americans but an invalid measure of intelligence for Mexican Indians.

Unlike reliability, the validation of an instrument occurs in many stages, over time, and with many different populations. Thus, a scale to assess problem solving that has been validated for college students will need further validation with middle-aged and the elderly before it is used with these individuals. Also, a scale that is valid is necessarily reliable, but the opposite may not be true. In other words, a scale may reliably or consistently measure the same phenomenon on repeated occasions, but it may not actually be measuring what it is intended to assess. Thus, validation of an instrument is extremely critical to assure some degree of accuracy in one's measure.

Where reliability is a measure of the random error of the measurement process itself, validity is a measure of *systematic, nonrandom error.* Nonrandom error refers to a systematic bias or to an error that occurs consistently. For example, a weight scale that systematically weighs individuals 5 lb heavier than an accurate scale is an example of systematic error and an invalid measurement. A measure that systematically introduces error indicates that it is a poor indicator of the underlying concept it is intended to represent. Validity is thus inversely related to the amount of systematic or nonrandom error present in an instrument: the smaller the degree of nonrandom error, the greater the validity of the instrument.

There are three basic types of validity that examine the degree to which an indicator represents its underlying concept: content, criterion, and construct.

Content Validity

Content validity is also sometimes referred to as face validity and is considered the most basic form by which an instrument is validated. This type of validity addresses the degree to which the indicator *seems* to reflect the basic content of the phenomenon or domain of interest. Ideally the steps to obtain content validity include (1) specification of the full domain of a concept through a thorough literature search and (2) adequate representation of domains through the construction of specific items.

If all domains are known to the investigator, a sampling of items to reflect each domain can occur. The problem with content validity is twofold. First, for most concepts of interest to health and human service professionals, there is no agreed upon acceptance of the full range of content for any particular concept. Take, for example, such concepts as poverty, depression, self-esteem, or function.

These concepts have been defined and conceptualized in many different ways. There is no unified understanding of the domains that constitute any one of these concepts. The second problem with this form of validity is that there is no agreed upon "objective" method for determining the extent to which a measure has attained an acceptable level of content validity. One way of obtaining some degree of assuredness of content validity is to submit constructed items or drafts of a scale for review by a panel of experts. This review process can be repeated until agreement has been obtained as to the validity of each item. However, this process does not yield a coefficient or other such statistical indicator to discern the degree of agreement or extent to which an instument is content valid.

Criterion Validity

This type of validity involves demonstrating a correlation or relationship between the measurement of interest and another instrument or standard that has been shown to be accurate. There are two types of criterion validity: concurrent and predictive . In concurrent validity, there is a known standardized instrument or other criterion that also measures the underlying concept of interest. It is not unusual for researchers to develop instrumentation to measure concepts, even if prior instrumentation already exists.

Suppose that you were interested in developing your own measure of self-esteem. To establish concurrent validity, you would administer your instrument along with an already accepted and validated instrument measuring the same concept to the same sample of individuals. The extent of agreement between the two measures, expressed as a correlation coefficient, would tell you if your scale was accurately measuring the same construct measured by the validated scale. The problem, however, is that this form of validity can be used only if another such criterion exists. Also, this form is only as good as the validity of the selected criterion.

Predictive Validity

Predictive validity is used when the purpose of the instrument is to predict or estimate the occurrence of a behavior or event. For example, this type of validity could be used to evaluate whether an instrument designed to assess risk of falls can predict the incident of falling. The degree of correspondence between the instrument and the criterion (a fall, or an established assessment of risk of falling) is indicated by a correlation coefficient.

Construct Validity

Construct validity represents the most complex and comprehensive form of validation. Construct validity is used when an investigator has developed a theoretical rational underlying the test instrument. The researcher then moves through different steps to evolve supporting evidence of the relationship of the test instrument to related and distinct variables. Construct validity is based not only on the direct and full measurement of a concept but also on the theoretical principles related to that concept. Therefore, the investigator who attempts construct validity must consider how the measurement of the selected concept relates to other indicators of that same phenomenon. The researcher who is developing a measure of depression would therefore base a set of expectations about the measurement outcome on sound theory about the nature of depression. If the instrument was a self-report of depression that was based in cognitive behavioral theory, the researcher would hypothesize that certain behaviors and thoughts, such as lethargy, agitation, appetite changes, self-effacing thoughts, and melancholia, would be frequently found in subjects who scored as clinically depressed. Thus, the relationship of the scale score to other expected relationships would be measured as an indicator of construct validity.

There are many complex approaches to construct validity. The validation of a scale is ongoing in which each form just discussed—content, concurrent, predictive, and construct—build on each other and occur in progression of one and the other.

Practical Considerations

In addition to the issues of reliability and validity, two other rather practical considerations influence the selection of a data collection technique. First, the nature of the research question dictates which method is appropriate. For example, if one is interested in surveying how professionals use theory in their practice, a closed-ended questionnaire would be developed. If one is interested in determining the accuracy of self-report of performance in self-care for older clients with serious mental illness, an interview and observation of performance may be appropriate.

Second, the nature of the sample also determines the appropriateness of the data collection procedure. Such factors as level of education, socioeconomic background, verbal ability, and cognitive status influence the selection of a data collection method. For example, college students are very familiar with a fixed-format type of self-administered questionnaire. However, this format may not be appropriate with the present cohort of elderly who may be visually impaired or unfamiliar with completing questionnaires. For this population, a face-to-face interview may be more appropriate. Questionnaires would also not be appropriate with children whose logical way of thinking cannot be captured in this format. The characteristics of a sample also determine the appropriate length of an interview or contact time with the investigator. For example, an individual with an attention deficit may not be able to attend to an interview that takes longer than 20 minutes.

INSTRUMENT CONSTRUCTION

Before concluding this chapter, let us examine some issues related to instrument construction. In health and human service research, the investigator often finds that measures have to be newly constructed to fit the study purpose. The development of a new measure in experimental-type research should always include a plan for testing its reliability and some level of validity (such as content, at the very minimum) to assure a level of accuracy and rigor to the study findings. Some investigators combine previously tested and established measures with newly constructed ones to test their relationships and relative strengths in the study.

The development of new instrumentation is, in itself, a specialty within the experimental-type research world. We have introduced you to the basic components of measurements and have given you some principles for ensuring that your items are reliable. However, constructing an instrument and its components is a major research task, and it is best to seek consultation.

The basic steps in instrument construction are:

- Review of the literature for relevant instrumentation and theory that underpins the development of new instrumentation
- Specification of the concept/construct to be operationalized into an instrument
- Conceptual definition of the full range and content of the concept
- Selection of an instrument format
- Translation of the concept into specific items/indicators with appropriate response categories
- Pilot testing of the instrument
- Subjection of the instrument to tests of reliability and validity

There are many different sources of error that interfere with the reliability and validity of an instrument. We list six considerations in the development of a data collection instrument and protocol:

1. The format or design of the instrument
2. Clarity of the instrument
3. Social desirability of the questions
4. Variation in administration
5. Situational contaminants
6. Response set biases (all "yes" or all "no" type of responses)

For more detailed information on instrument construction, we refer you to the bibliography in which we have listed texts that are specifically devoted to that topic.

ADMINISTERING THE INSTRUMENT

In many cases, as in small studies or in studies using written response instruments, administering the instrumentation is a function of distributing a paper and pencil assessment to subjects or directly observing the phenomena of interest. However, it is not unusual to engage in a project where more than one researcher is obtaining information. In projects of this sort, the influence of the data collector can potentially confound the findings. Let us consider how.

Suppose you and another investigator are studying perceived quality of life in a nursing home environment. You both have a set of questions with a closed-ended response set to be answered by the subjects. One investigator begins the interview with a thank you statement and then begins the questions, whereas the other interviewer discusses a respondent's poor health status before beginning the interview schedule. In your data analysis, you find that the subjects interviewed by investigator 1 tend to be more satisfied with their lives than those interviewed by investigator 2. Could it be possible that investigator 2 influenced the subjects by discussing potentially depressing issues before the administration of the instrumentation? This scenario highlights the importance of training for data collectors. Training to assure that data collection procedures are the same for all data collection action processes is essential to reduce bias that can be introduced by investigator difference. That is, establishing a strict protocol for administering an interview and training interviewers in that protocol assure standardization of procedures such that "objective measurement" can occur.

SUMMARY

Data collection methods for experimental-type design vary along several dimensions that include the degree of:

- Structure (open or closed)
- Quantifiability (level of measurement)
- Objectivity (reliability and validity)

The purpose of data collection is to assure minimum researcher obtrusiveness through the systematic and consistent application of procedures (in other words, training). The researcher tries to minimize systematic error through validation and random error through reliability testing. Other procedures are developed to assure consistency in obtaining data, such as:

- Establishing testing procedures and protocols
- Establishing when and how subjects are contacted
- Establishing a script by which to describe the study

Reliability and validity represent two critical considerations when a data collection instrument or indicator is selected. When experimental-type research is conducted, one's first preference is the selection of instruments or indicators that have demonstrated reliability and validity for the specific populations or phenomena one wishes to study. However, in health and human service research, as in other substantive or content areas such as gerontology, there are a limited number of developed instruments to measure the range of relevant researchable concepts. Because of the lack of reliable and valid instruments, some professional organizations, such as the American Occupation Therapy Foundation, have placed a high priority on funding research that are directed at the development of instruments.

EXERCISES

1. Develop at least five questions on a Likert scale to measure job satisfaction among your peers. Ask at least three of your peers to respond to the questions. Also ask each to indicate the clarity of the questions. Do you need to revise the questions? Did the questions appear to measure what you intended it to?

2. Using job satisfaction as your construct, now develop a Gutman scale and a semantic differential scale. What are the advantages and disadvantages of each? Ask your peers to respond to these questions as well, and compare the responses.

3. With a peer, observe a child in a playground and attempt to determine his or her age independently. Record the criteria that you used to make a judgment. Compare your impressions with your peers to determine the extent of inter-rater reliability between the two of you. Now repeat the process using the same criteria to judge age.

REFERENCES

1. Kerlinger F: *Foundations of behavioral research,* New York, 1973, Holt, Rinehart & Winston.
2. Miller D: *Handbook of research design and social measurement,* Newbury Park, Calif, 1991, Sage.
3. Spanier GB: Measuring dyadic adjustment: new scales for assessing the quality of marriage and similar dyads, *J Marriage Fam* 38:15–28, 1976.
4. Zung WK: A rating instrument for anxiety disorders, *Psychosomatics* 12:371–379, 1971.
5. Olson DH, Portner J, Lavee Y: FACES-III. In *Family social science,* Minneapolis, 1985, University of Minnesota.
6. Spielberger CD, Jacobs G, Russel S, et al: Assessment of anxiety: the state-trait anxiety scale, *Adv Personality* 2:159–187, 1983.

7. Stiles WB, Snow JS: Counseling session impact as seen by novice counselors and their clients, *J Counseling Psychol* 31:3–12, 1984.
8. Radloff LS: The CES-D scale: a self report depression scale for research in the general population, *Appl Psychol Measurement* 1:385–401, 1977.
9. DeVellis RF: *Scale development: theory and applications,* Newbury Park, Calif, 1991, Sage.
10. Stouffer SA: *Communism, conformity and civil liberties,* New York, 1955, Doubleday, pp 263–265.
11. Osgood CE, Suci GJ, Tannenbaum PH: *The measurement of meaning,* Urbana, Ill, 1957, University of Illinois Press.
12. Wechsler D: *The measurement of adult intelligence,* ed 4, New York, 1958, Psychological Corp.

15

Data Gathering in Naturalistic Inquiry

- **Investigator Involvement**
- **Gathering Information and Analysis**
- **Prolonged Engagement in the Field**
- **Multiple Data Collection Approaches**
- **The Process**
- **Accuracy in Data Collection**
- **Summary**
- **Exercises**

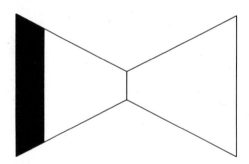

GATHERING INFORMATION
• Investigator-informant relationship

Let us shift worlds and perspectives again and examine the actions of researchers engaged in gathering information within a naturalistic-type of inquiry. In this tradition, knowledge is "mind dependent" in that it emerges from and is embedded in the shared understandings and interactions of individuals. Therefore, gathering information involves a set of investigative actions that are quite divergent from experimental-type research in purpose, approach, and process. The overall purpose of gathering information is to discover or reveal the multiple and diverse perspectives and patterns that structure field experience. The investigator gathers sufficient information that leads to description, discovery, understanding, and explanation of the rich mosaic of daily life experiences or the context of the study. As you may recall from previous chapters, there are many distinct perspectives along the continuum of naturalistic inquiry. Consequently, researchers use many different approaches to gathering information to accomplish this purpose. For example, in semiotics,[1] the focus of data collection is on discourse, symbol, and context, whereas in ethnography,[2] the focus of data collection may be much broader and may include not only verbal interchange but behavioral patterns and objects as well. Because of this variation, this chapter presents a discussion of the basic principles that guide gathering information and that tend to be shared by researchers along the naturalistic continuum. We refer you to other books, listed in the bibliography, that discuss specific design strategies in naturalistic inquiry and their data collection approaches.

Four basic principles in gathering information tend to be shared by researchers along the continuum of naturalistic designs: investigator involvement, the interactive process of gathering information and analysis, prolonged engagement in the field, and use of multiple data collection strategies. Let us examine each guiding principle.

INVESTIGATOR INVOLVEMENT

In naturalistic inquiry, the primary instrument for gathering information is the investigator. As Fetterman[3] describes:

> The ethnographer is a human instrument. Relying on all its senses, thoughts, and feelings, the human instrument is a most sensitive and perceptive data-gathering tool [p 92].

The investigator, as a data-gathering instrument, is a logical extension of the primary contention of naturalistic inquiry. This position maintains that the only way to "know" about the "lived" experiences of individuals is to become intimately involved and familiar with the life situations of those who are studied. As the primary data-gathering instrument, the researcher strives for intimate familiarity with the people in a study. By engaging in an active learning process through watching, observing, participating, interviewing, and reviewing materials, the investigator enters the life field of others and tries to uncover what it means to live that experience.

GATHERING INFORMATION AND ANALYSIS

Another shared principle by researchers along the continuum of naturalistic inquiry is that gathering information involves ongoing analysis of the information that is collected. Fieldwork is a dynamic process in which data collection and analysis are linked so that one action informs the other. The investigator thinks about and evaluates information as part of being immersed in the field and actively observing, taping interviews, or participating in activities. In Chapter 18, we discuss the specific forms of analyses that occur in tandem with these data collection actions. Initial analyses suggest the "who, what, when, and where" aspects that define further data collection efforts. Ongoing data gathering, analysis, and more data gathering can be viewed as a spiraling process. In this spiraling process, as analysis and data collection become refined, the investigator moves from a broad conception of field reality to a more in-depth, richer, or "thick"[4] understanding and interpretation of multiple field "realities."

PROLONGED ENGAGEMENT IN THE FIELD

Understanding a natural setting involves studying it in all its complexity. The investigator must therefore be sufficiently immersed in the setting and spend sufficient time in the field to assure that a rich, or thick, and adequate description of that field can emerge. Prolonged engagement or selection of data gathering approaches to assure obtaining in-depth understandings is critical to studies along this continuum. It is the investigator who determines when sufficient description has been obtained. The extent of "immersion," or amount of time spent in data collection, varies, depending on the initial query and availability of necessary resources and obviously investigator time.

MULTIPLE DATA COLLECTION APPROACHES

In most studies along the continuum of naturalistic inquiry, the investigator draws on and combines a variety of data collection strategies to obtain a rich description and deep understanding. There are many different data-gathering strategies that we discuss in more detail. These data collection approaches are not necessarily planned or staged before entering the field. Rather, the investigator makes decisions as the research progresses regarding the timing of the introduction of any one particular strategy. Decisions, for example, as to when, how, and whom to observe and interview emerge from the ongoing data collection–analysis–data collection–reanalysis process discussed earlier.

Let us summarize our discussion thus far. The hallmark of gathering information for the naturalistic researcher is that this action process occurs in a natural context. These actions are designed to examine that context and all that occurs within it. The critical components of these actions are that it is the investigator who is actively involved in the context over a prolonged time period to assure sufficient immersion in the phenomena under study. Multiple strategies are used to obtain information in an ongoing data collection and analytic integrated process. These basic components or guiding principles in data gathering yield interpretations and understandings of phenomena as they exist embedded in specific contexts and complex interrelationships. This complexity is preserved by the nature of the data gathering process.

THE PROCESS

So what actually is the process of doing fieldwork or gathering information? Although the process has been variably defined, we borrow the structure suggested by Shaffir and Stebbins[5] in *Experiencing Fieldwork* and view it as involving four interrelated parts or considerations: context selection and "getting in the field"; "learning the ropes," or obtaining meanings of the setting; "maintaining relations"; and then "leaving the field." Although these actions may appear to represent distinct phases or research steps, the process is truly integrated in which each component overlaps and shapes the other. Here we examine context selection and learning the ropes and refer you to other texts that address the specific issues investigators confront in maintaining fieldwork relationships and then leaving the field.

Context Selection and Getting in the Field

The first decision about data collection in naturalistic research is where or in what context information should be obtained. Because naturalistic research focuses on the natural research context, as well as what happens within it, selection of a context is a very important factor in data gathering. Context selection may sound simple, but several important considerations arise. First, the researcher must bound the study by depicting a conceptual or locational context. (Remember Chapter 12?)

Second, the specification of a context must be practical or realistic to study. However, in bounding a study to a particular context, the researcher must be careful not to limit the focus and thus the opportunity to obtain a full picture of the phenomenon under investigation. For example, Hasselkus[6] examined the use of therapeutic activities in one adult day-care center in the Midwest section of the United States. Perhaps if her context had included adult day-care centers in different locales or had been defined as client-staff interactions, different interpretations of the therapeutic modalities may have emerged.

Let us consider the investigator who is examining the functional capacities of persons with traumatic brain injury. If the context for the study is functional capacity in a clinical setting, the informant may be maximally functional in that setting. The investigator who then neglects to expand the context to other environments may not reveal the functional incapacity that persons with traumatic brain injury often exhibit in less structured settings.

Once a context has been identified, the investigator must gain access to the field. This is not necessarily as simple as it sounds. It is a critical aspect of naturalistic inquiry and requires strategizing, negotiation, and ongoing renegotiation. If the investigator is already part of the natural context to be experienced (an insider), access is not necessarily a problem but reflexivity is. In other words, the insider must understand how knowledge and previous experiences of the research context influence the information gathered and how it is interpreted. Consider, for example, the study of the culture of a nursing home by DePoy and Archer.[7] Archer, in this research effort, was employed as an occupational therapist in the nursing home in which the study was conducted. She also functioned as the primary data collector. Because of her familiarity with the setting, she had no difficulty gaining entry and conducting observations. She was a familiar and trusted member of the context. Although this was an advantage, Archer had to work hard to ensure that she did not impose her previously conceived notions of that setting on the information that she gained and that she understood how her insider status influenced her data collection and analytic actions.

If the investigator is an "outsider," access must be obtained. Ethnographers in particular have extensively discussed this aspect of naturalistic inquiry and have highlighted that the way in which access and entry are obtained often influences the entire course of the research process. Gaining access involves obtaining permission to enter and become part of a social or cultural setting. Investigators use many different types of strategies, depending on the nature of the context and their initial relationship to it. Formal introduction to an informant by another, slowly building rapport through participation, and gift giving are just a few of the techniques used by investigators to enter a field.

A component of the issue of access is that of disclosure or concealment. The researcher who conducts a naturalistic inquiry must decide on the extent to which the research activity is revealed to those being observed or interviewed. Although disclosure may occur initially to those informants who provide initial access for the investigator, the specifics of the research activity may remain vague or unknown to others who are either observed or interviewed at different points in time in the course of fieldwork. The degree and extent of overt or covert research activities vary within any one type of inquiry and from study to study and have important ethical concerns for researchers along this continuum.

The other aspect of disclosure is that of assuring confidentiality of information. For example, Stearns[8] discusses this issue for physicians in her descriptive study of the way in which HMOs structure professional relationships between physician and patient:

> Only a few physicians refused to allow me to tape the interview. However, many expressed concerns that the interview remain confidential.

This concern as well as their hesitancy to be interviewed meant that I had to be especially careful not to reveal who I had interviewed or what they had said. Confidentiality of physician (and other) responses was thus of great concern in the research [p 331].

Whether the investigator is initially an insider or outsider, his or her presence in and of itself may change the nature of that field and how people present and live their experiences. The reactive effects of the investigator's presence on the phenomena being observed is a major methodologic dilemma for this type of research.

Liebow,[9] in the study of the culture of black men who frequented an urban street corner that we have previously discussed, describes the effect that his presence had initially in the culture to which he attempted to gain access:

All in all, I felt I was making steady progress. There was still plenty of suspicion and mistrust, however. At least two men who hung around Carry Out—one of them the local numbers man—had seen me dozens of times in close quarters but kept their distance and I kept mine [pp 242–243].

"Learning the Ropes," or Obtaining Meanings of the Setting

Once initial entry into the context has been made, gaining access to different informants and types of observations remains an ongoing process that involves continued negotiation and renegotiation with members of a site. This ongoing access process is part of the active work of doing fieldwork, which has been variably referred to as "learning the ropes,"[5] "obtaining meanings," or obtaining an "intimate familiarity"[10] with the setting. Although this process may appear as if the investigator is "just hanging out," a range of data-gathering strategies, which include watching and listening, asking questions, and examining materials, are brought into play to learn meanings within a setting. These strategies are purposely chosen, combined, and integrated as a consequence of the unfolding process of learning about the field. The choice of strategy depends on the ebb and flow of the fieldwork situation and the particular issue on which the investigator is focused at the moment. Each strategy is designed to move the investigator from being an outsider to an insider or member of the group capable of understanding and experiencing the meanings of symbols, language, and behavior within the setting.

Researchers along the naturalistic continuum believe that it is through people using their own words and meanings and experiencing their experiences that the investigator comes to know the context. Thus, as you will see, the three data-gathering strategies have a different purpose and structure when used in naturalistic inquiry than when used in designs along the experimental-type continuum.

Watching and Listening

As we discussed in Chapter 14, the process of watching and listening is often referred to as "observation." In naturalistic research, observation occurs within the natural context, or the field. In this data collection strategy the investigator can be either a passive observer or a participant. Lofland and Lofland[10] define participant observation as: "the process in which the investigator establishes and sustains a many-sided and relatively long term relationship with a human association in its natural setting for the purpose of developing a scientific understanding of that association."

In participant observation, the investigator is not a passive observer but actively engages in the research context to come to know about it. The investigator begins with broad observations to describe what is seen and then narrows the focus to discover the meaning of what has been described. In descriptive observation, it is often helpful to think of yourself as a video camera and to record what the video camera would see as it scans the boundaries of the research. This technique reminds the researcher that description, not interpretation, is the first step in participant observation.

Participant observation has many things in common with what we do in newly encountered social situations. All of us participate and observe in many social situations. However, as a researcher engaged in participant observation, one seeks to become explicitly aware of the way things are. The researcher remains introspective and examines the "self" in the situation and his or her experiences as both the insider participating in the event and the outsider observing the actions.

Krefting[11] describes why and how she used participant observation for her study on individuals with brain injury and their family members:

> The second fieldwork strategy was participant observation in which I, as the researcher, was both engaging in and observing the scene. In participant observation I recorded my own actions, the behaviors of others, and aspects of the sociocultural situation. Participant observation is a particularly valuable strategy in cases of head injury, because what people say is not always reflected in what they do. In this study, I engaged in participant observation in three ways. First, I attended various meetings of family support groups and related social events. Second, I attended treatment sessions, either at schools or at day treatment centers. Third, I spent time with head-injured people and their families: together we went shopping, ate in restaurants, and visited neighbors. I made no attempt to interview professionals: my goal was to experience life from the perspective of a head-injured person [p 69].

Gubrium[12] summarizes the advantages of this approach:

> The goal is to wind up with a depiction of "the way it is," as it were, in a particular setting. . . .we can represent the meanings of a setting in terms

more relevant to our subjects than other methods permit. "I've been there," the participant observer likes to put it, "seen what actually happens, and this is the way it is" [p 140].

Some types of designs or fieldwork situations may warrant less participation in the observation process. Nonparticipatory observation can be used to obtain an understanding of a natural context without the influence of the observer. For example, Runcie[13] suggests that a student may observe the patterns of activity in a public location without interfering in the action as a basis for understanding the time and action cycle that occurs. To then contrast participation with nonparticipation, Runcie suggests that the observer make himself or herself known to the persons in the location in some way and observe the differences in the action. The nonparticipant observer may record field notes while watching and listening or may choose to log observations after leaving the scene. Each has its advantages. Recording while watching may call attention to the observer, who would then be more participatory in the setting. However, waiting to recall descriptions makes us rely too much on our memory.

During the fieldwork process, watching and listening tend to move from broad, descriptive observation to a narrower focus in which the observer looks not only for more descriptive data but also for specific understandings of the meanings of what he or she has observed. The degree of participation in the context and the extent of immersion vary throughout fieldwork and is based on the nature of the inquiry, access to the setting, and practical limitations.

Asking

Asking, or interviewing, as a method of data collection, is another critical strategy used by the naturalistic researcher. The investigator may use an asking strategy to obtain an example of a social interchange, to clarify or verify the accuracy of observations, or to obtain the informant's experience with and view of the phenomenon under study. Asking, unlike observation, involves direct contact with persons who are capable of providing information. Thus, to collect data by asking, one must establish a relationship appropriate to the level of involvement with the informant.

Asking can take many forms in naturalistic research from an informal, open-ended conversation to focused or long in-depth interview.[14] The timing of the interview in the research process, the purpose of the social interchange, and the nature of the relationship with the informant all will determine the type of asking that the investigator will choose.

The most common form of asking in naturalistic design is unstructured, open-ended interview. Read how Krefting[11] describes her use of interviewing as one fieldwork strategy in her ethnography of individuals with brain injury and their family members:

> The first was extensive nonstructured interviews with disabled persons, their families, and their friends. All interviews were taped with informed

consent. I conducted over 80 interviews, which ranged from 1 to 4 hours [p 69].

In ethnography, this type of interviewing is frequently used initially. The interview may consist of informal conversation, recorded by the investigator, or may consist of face-to-face twosomes or small groups in which the investigator sets the context for the interview and the informants offer their knowledge.

Among the most common face-to-face interviews are ethnographic interview and life history. In ethnographic interview, the investigator seeks to understand culture through talking with insiders in the culture. The investigator not only attends to the content of the interview but also examines the structural and symbolic elements of the social exchange between the investigator and the informant. In other words, contemporary ethnographic interview acknowledges that meaning may be changed and made within the context of the interview itself.[15-17]

Life history is a form of ethnographic interview that chronicles an individual's life within a social context. The investigator elicits information not only on the important events, or "turnings"[18] in an individual's life, but on the meaning of those events within the sociocultural context. This form of interviewing can also be an independent design structure.

Common forms of group asking include focus groups and interviews with social units, such as families that contain more than one individual. In a focus group,[19] the investigator sets the parameters for the conversation. The participants, usually a group of five to ten individuals, then address the topic guided by the investigator. This approach is used when it is believed that the interactions and group discussion will yield more meaningful understanding than single, independent interviews. Interviews with small social units are also useful in observing group dynamics and symbolic meanings exchanged among group members. In cultural descriptions, asking social units can be most valuable in understanding the structural groups of a society. This form of interviewing can also be an independent design structure and was presented in Chapter 11.

Although use of structured questionnaires has been used primarily in experimental-type designs, it is becoming more popular in naturalistic research. Investigators have found that open-ended, or even semistructured, questionnaires provide organization and boundaries for the voluminous data that are generated in naturalistic asking. These more structured forms of asking are very purposeful and are introduced once the investigator is in the field, has developed rapport, and learned the ropes to know what to ask and how to ask it.

Another use of the interview is to verify the investigator's conclusions to determine if they are representative of the perspective and experience of the phenomena and persons observed.

Let us examine one strategy of asking that may be pursued in a field study. The four interrelated components of this particular strategy involve:

- Access
- Description
- Focusing
- Verification

Asking begins with access. The investigator selects or meets an individual or individuals who can best illuminate the inquiry. For example, in her study of homeless, middle-aged women, Butler[20] entered the field with broad criteria for selecting informants, which included those who were willing to talk to her over a 3-month period and those who were willing and able to articulate their experience. Then, much of her time was spent hanging out and establishing rapport with informants. Once a level of rapport and confidence was established, she was able to apply more purposeful and structured interviewing approaches.

In the study by Liebow,[9] access was initially gained through hanging out on the street corner where men were likely to congregate. Liebow describes his first entree into the social scene as one that occurred serendipitously. On his first day in the field, he noticed a commotion and immediately went to the site where it was occurring. As a result of being there, Liebow was able to introduce himself into the culture of the street corner.

Once access is obtained, the initial goal of asking is to describe. Asking begins with broad questions, which then become more focused as trends, recurrent patterns, and themes emerge. Frequently the investigator begins the asking process with a broad probe such as "tell me about. . . ." In the study by Liebow,[9] asking began informally when he invited a man whom he had met on the street corner to have coffee. Liebow[9] clarified his task as a researcher and "sat at the bar for several hours talking over coffee" [p 238].

Based on emerging descriptive knowledge, the researcher's asking becomes more focused and probing. The asking process involves ongoing verification of impressions, descriptions, and understandings.

The interview process may occur in one session, several sessions, or many sessions and is often combined with observation and review of existing materials. Of no surprise, the decision as to length and frequency of interview is dependent on the query, the purpose of the interviews, and practical limitations influencing the study.

Examining Materials

The examination of materials such as records, diaries, journals, and letters has become an essential data collection strategy in contemporary naturalistic research. This is particularly the case in health and human service research where written records, such as charts, progress notes, or minutes, are routinely kept.

Similar to watching, listening, and asking, review of materials begins with a broad examination and then a more focused evaluation to explore recurring themes and patterns. Knowledge about what materials are available may occur at any point in the research process. It is not unusual to plan to examine objects and records at the beginning of a project. Suppose, for example, you were interested in determining the extent to which a behavioral program for persons with mental retardation was improving social behavior in sheltered work environments. Although you might observe workplace behavior, another effective strategy would be to obtain information on "unobserved behavior," that which was not observed by the researcher. For example, you might examine progress notes written by the sheltered workshop staff. These notes would reveal behavioral patterns and the context in which behavior was occurring as perceived by the staff. Their perceptions would then be analyzed in light of your own observations and perceptions and the other forms of information that are obtained.

To access written documents, the investigator must obtain appropriate consent, even if he or she is an insider. For nonobtrusive data collection, consent may not be desirable, because a primary aim of collecting the information using these sources is to keep the research role covert. Consider the researcher who is interested in determining substance use patterns in teenagers in a small town setting. Examining the contents of the trash for alcohol containers or other paraphernalia that could shed light on substance abuse patterns may be the most accurate way to estimate alcohol consumption in a group, because it is unlikely that youth will readily report their behavior.

Krefting[11] describes the range of materials she used as a data collection strategy:

> The third fieldwork strategy was a review of pertinent nontraditional documents. These included human interest stories from local newspapers, newsletters from the National Head Injury Foundation and state organizations, brochures describing new treatment programs, and personal documents provided by informants. These personal documents were perhaps the richest source of data. One such document was a diary that had been kept for 2 years. Used as a memory prompt, it detailed all the major events in the informant's life. I also read several testimonials, or what people called head injury stories, prepared by the informants, and even the lyrics of a song describing the head injury experience [p 69].

To summarize, the action process of data collection for the naturalistic researcher involves the use and combination of three basic methods to gather information: watching and listening, asking, and examining materials. Watching and listening occurs in the form of observation and ranges on a continuum from full to no participation in the context that is being observed. Asking also ranges in structure with interviewing as the primary asking process in naturalistic design. Examination of materials occurs in an inductive way and is intended to reveal patterns related to the phenomena under study. Naturalistic data collection occurs along with analysis and moves from broad information gathering to more focused collection of information. The combination of these approaches is critical to obtain breadth and depth of analysis.

As descriptive data are collected to illuminate the answer to "what, where, and when," the next step in the process is to focus data collection efforts based on the ongoing analytic process. The process of information collection continues with "why" and "how" questions. In ethnography, this effort is called thick description,[4] or focused observation,[21] in which meanings are sought.

Focused observation hones in on examining the patterns and themes that emerge from descriptive information. The investigator may choose to examine one pattern at a time or may focus on more than one. The investigator narrows the scope of inquiry to themes and the interaction between patterns and themes while remaining open to further discovery.

In general, collection strategies provide information to answer the questions: where, who, how, when, and why. Answers to these five questions yield description, understanding of the context and the occurrences within the context, and the timing.

Recording Data

An essential component of learning the ropes and watching, listening, asking, and examining materials is recording information as it is obtained. Data recordings, although described in a multitude of ways in the qualitative research literature, have two basic components: recording of events, observations, and occurrences; and recordings of impressions of events, observations, and occurrences. There are numerous formats suggested in the literature for recording descriptive field notes. According to Spradley,[21] a matrix design can be used to look at the interaction of persons, places, and objects. However, other formats are equally as useful. Lofland and Lofland[10] suggest recording description and bracketing impressions and interpretation within the context of the description.

We have found it useful to use a stenographer's pad on which the left side is used to describe and the right is used for the investigator's thoughts, impressions, and questions. The separation of description from interpretation is also useful for analyzing and guiding the next step of focused observations.

Krefting[11] describes the four approaches she used in recording data:

> First, I recorded interviews on audiotape. I transcribed these tapes into the computer immediately after the interview had taken place. Second, I prepared field notes using the short-note technique of recording data immediately after concluding interviews. These notes contained my perceptions regarding the interviews and the activities observed. General impressions of interviews, particularly of the participants' moods, were noted, as was interactive data. Third, I reviewed relevant documents. Fourth, I kept a field log, or diary, that recorded the thoughts, feelings, ideas

and hypotheses generated by my contact with informants. This log also contained my questions, problems, and frustrations concerning the overall research process [p 70].

Audiotaping, photographs, and videotaping are also useful tools that can be used to supplement diary or note taking.

ACCURACY IN DATA COLLECTION

At this point, you may be asking how accuracy or trustworthiness of data collection is assured in naturalistic design. In Chapter 18, we discuss a number of approaches or techniques investigators use to enhance the trustworthiness of their data collection efforts that are integrated in the analytic process. Here we would like to bring to your attention two basic approaches to enhance accuracy: the use of multiple data gatherers and triangulation.

Multiple Data Gatherers

The first technique is the use of two or more investigators in the data-gathering and analytic process. The old adage "two heads are better than one" is also true in naturalistic inquiry. If possible, it is useful for two or more investigators to independently observe and record their own field notes. This technique checks the accuracy of the observations in that more than one pair of eyes and ears are examining and recording the same context. Careful training of both observers is essential to ensure that skilled recording and reporting are accomplished by each observer and that all investigators understand the purpose and intent of the study.

Consider an example of an investigator who is studying the culture of a group of hospitalized adolescents. The purpose of such a study may be to determine the norms of the culture as it emerges within the boundaries of the hospital unit. Using the technique of multiple observers, two or more investigators would keep field notes independent of one another and initially analyze these data independently as well. The investigators would then meet to compare notes and impressions and reconcile any differences through in-depth discourse about the data set. As you can see, not only are multiple observers used, but accurate data collection is also enhanced by multiple analyzers.

Triangulation

The second action process that increases the accuracy of information gathering is triangulation of data collection methods.[22] Triangulation of data collection refers to using more than one strategy to collect information bearing on the same phenomenon. The use of observation with interview, examination of materials, or both is characteristic of triangulation in traditional naturalistic research. More recently, quantitative measures and analytic techniques such as content analysis have also been introduced to provide meaning of a phenomenon from a different angle. In the hypothetical example of an adolescent culture in a hospital setting, impressions about self-esteem within the culture, obtained through observation, may be further verified with interview and self-esteem testing techniques.

SUMMARY

The main purpose of the action process of data collection in naturalistic inquiry is to obtain information that incrementally leads to the investigator's ability to reveal a story, a set of descriptive principles or understandings, hypotheses, or theories. Each piece of information is a building block that the investigator inductively collects, analyzes, and puts together to accomplish one or more of the purposes just stated. In naturalistic research, data collection and analysis go hand in hand, and data collection continues to unfold dynamically as analysis reveals further direction for information gathering. The investigator is the main vehicle for data collection and assurance of accuracy. To enhance accuracy of data collection, multiple observers and triangulation of collection methods are two action processes often used by naturalistic investigators.

EXERCISES

1. Select a public place, such as a shopping mall, outpatient clinic, or public transportation station. Use nonparticipatory observation and then participatory observation to determine behavioral patterns that characterize the human experience in that setting. Compare your experiences using both forms of observation and what you learned from each.

2. Conduct an open-ended interview with a colleague to obtain an understanding of career choices and their career path. What other information gathering techniques could you use to understand this phenomenon?

3. Select a research article that uses naturalistic inquiry. Identify the information-gathering techniques used and determine the ethical issues involved for each technique.

REFERENCES

1. Hodge R, Kress G: *Social semiotics,* Ithaca, NY, 1988, Cornell University Press.
2. Agar MH: *The professional stranger: an informal introduction to ethnography,* New York, 1980, Academic Press.

3. Fetterman DL: A walk through the wilderness: learning to find your way. In Shaffir W, Stebbins R, editors: *Experiencing fieldwork: an inside view of qualitative research*, Newbury Park, Calif, 1991, Sage.

4. Geertz C: *Interpretation of cultures*, New York, 1973, Basic Books.

5. Shaffir W, Stebbins R, editors: *Experiencing fieldwork: an inside view of qualitative research*, Newbury Park, Calif, 1991, Sage.

6. Hasselkus BR: The meaning of activity: day care for persons with Alzheimer disease, *Am J Occup Ther* 46:199–206, 1992.

7. DePoy E, Archer L: The meaning of quality of life to nursing home residents: a naturalistic investigation. *Topics Geriatr Rehabil* 7:64–74, 1992.

8. Stearns CA: Physicians in restraints: HMO gatekeepers and their perceptions of demanding patients, *Qualitative Health Res* 1:326–348, 1991.

9. Liebow E: *Tally's corner*, Boston, 1967, Little, Brown.

10. Lofland J, Lofland L: *Analyzing social settings: a guide to qualitative research*, Belmont, Calif, 1984, Wadsworth.

11. Krefting L: Reintegration into the community after head injury: the results of an ethnographic study, *Occup Ther J Res* 9:67–83, 1989.

12. Gubrium J: Recognizing and analyzing local cultures. In Shaffir W, Stebbins R, editors: *Experiencing Fieldwork: An inside view of qualitative research*. Newbury Park, Calif, 1991, Sage, pp 131–142.

13. Runcie JF: *Experiencing social research*, Homewood, Ill, 1976, Dorsey Press.

14. McCraken G: *The long interview*, Newbury Park, Calif, 1988, Sage.

15. Spradley JP: *The ethnographic interview*, New York, 1979, Holt, Rinehart & Winston.

16. Van Maanen J, editor: *Qualitative methodology*, Newbury Park, Calif, 1983, Sage.

17. Merton RK, Fiske M, Kendall P: *The focused interview: a manual of problems and procedures*, New York, 1956, Free Press.

18. Langness LL, Frank G: *Lives: an anthropological approach to biography*, Novato, Calif, 1981, Chandler & Sharp.

19. Krueger R: *Focus groups: a practical guide for applied research*, Newbury Park, Calif, 1988, Sage.

20. Butler S: *Homeless women*, dissertation, University of Washington, Seattle, 1991.

21. Spradley JP: *Participant observation*, New York, 1980, Holt, Rinehart & Winston.

22. Miles MB, Huberman AM: *Qualitative data analysis: a sourcebook of new methods*, Newbury Park, Calif, 1984, Sage.

16

Analyzing Information

- **Experimental-Type Research Approach to Analysis**
- **Analysis in Naturalistic Inquiry**

We have thus far discussed the different strategies researchers use to collect information. But what do you do with the massive amount of information that you obtain? As in each research phase, analyzing information is approached quite differently for the experimental-type researcher and the researcher using a naturalistic design. In this chapter, we discuss what constitutes data for each continuum and how each handles and prepares the information for analysis.

EXPERIMENTAL-TYPE RESEARCH APPROACH TO ANALYSIS

If you are conducting experimental-type research, you are primarily interested in obtaining a quantitative understanding of a phenomenon. Your data are numeric, and analysis of these numbers is an action that is performed after the completion of data collection. At the conclusion of your data collection effort, you have numeric responses to your questionnaires, interviews, or observations. Your next step is to organize the information to facilitate the statistical manipulation of the data. Even in small pilot studies, the amount of information obtained is so massive, that the assistance of computer technology is usually required. Thus, the primary action process of preparing data for analysis on the experimental-type continuum is the entry of the data into a computer. Each numeric response generated by your subjects may be entered in the computer using a data entry program. Although a variety of programs are available, each is basically the same. But first, before entering data, the researcher needs to assign each variable on the data collection instrument to a unique location in the computer program. Most researchers use the convention of placing each observation along an 80-column format, as displayed in Table 16–1. Usually subject identification numbers are located in the first three to five

TABLE 16–1.

Example of Raw Data Set

10291	17823123338282883842343435012931242224
10291	2221132000021231
10291	301924302 010212212736572524761
20822	15721222112729318883984793898426547446
20822	2634152346520012354365415142534256342623423
28822	368013223 654272342546546545212
12311	22342123523642465256415642653462635
12311	1812235552516617263555274354413254231
12311	381726357 647652763674500000000
21071	18321122112535645625334132432434234323
21071	21222424424242442431313322525252525251111122344
21072	318723648 7234619872346871112333

columns of a line or row. If there is more than one line of data, that subject identification number is repeated for each row.

In Table 16–1, each number represents a response to a data collection instrument, in this case a closed-ended questionnaire. These raw numbers are referred to as a *data set* and form the foundation from which the experimental-type researcher performs statistical manipulations. A data set merely consists of all the numbers obtained through data collection, which are then organized according to the 80-column format. This data set, once entered, becomes the basis from which analytic procedures are performed.

To remember the line and column placement of each observation, the researcher records the information in a *codebook*. The excerpt of a codebook displayed in Table 16–2 indicates one style of developing this type of record. This example explains the location of variables for the data set displayed in Table 16–1. For example, according to the codebook, the first column on each line in Table 16–1 refers to a subject's group assignment (experimental or control group). Columns 02 through 04 refer to the personal identification number. This is a unique identifier for each participant in the study. Column 05 indicates the respondents gender; column 06 is blank and serves just to visually separate the subject's identification from data responses. Some investigators insert a blank after every 10 numeric scores to provide visual relief during data cleaning (discussed below). Everyone develops his or her own style and conventions. Column 07 refers to the specific line of data. For example, in the data set shown in Table 16–1, there are three lines (1, 2, and 3) of data. Columns 08 to 09 record the subject's age and so forth.

See if you can read the data set and codebook. First, consult the codebook in Table 16–2 to find the line and column in which the variable "age" is recorded. Then, in Table 16–1, the data set, find subject identification number 20822. Now what is the age of this subject? If you found this subject's age to be 57 on line 1, columns 08 to 09, you followed the codebook and data set correctly.

Before the data set is actually manipulated, it must be *cleaned.* Cleaning the data is an action process whereby the investigator checks the inputted data set to ensure that all data have been accurately represented and recorded. This process is essential to determine the extent of missing information, to assure that interviews and responses have been coded correctly and consistently, and to assure that no errors were made in entering the numbers in the computer. Just think if the age for subject 20822 was entered in the computer as 27 because of a data entry error. This error would certainly skew our understanding of the mean age and range of ages of the sample in our study. In data collection, we talked about the possibilities of random and systematic error. Data entry represents yet another possible source for random error.

Researchers use many techniques to clean data. Computing frequencies of responses for each variable once all data have been entered and then visually checking them can reveal possible errors such as improper recording or data entry. For example, suppose you have a response set from only 1 to 4 for your items. If you compute frequencies and see that one item has a response of 5, you know that datum is incorrect and needs to be rechecked and corrected. It is critical that a careful and systematic examination of the raw data set and initial frequencies of key variables be conducted. Although this is often a time-consuming task, it increases accuracy and confidence in your data set. No

TABLE 16–2.

Example of Codebook

Column No.	Line No.	Item	Variable Name
01	1	Experimental Assignment 1 = Experimental 2 = Control	Exper
02–04	1	Personal identification number	Period
05	1	Sex 1 = Female 2 = Male	Sex
06	1	Blank	
07	1	Line location	Line
08–09	1	Age	Age
10–11	1	Total anxiety score (Range 00–40 low score = high anxiety)	Anxiety
12	1	In the past week, how tense have you been? 1 = Always 2 = Occasionally 3 = Somewhat 4 = Rarely 5 = Never	Tense

statistical manipulations can be conducted until the data set has been thoroughly cleaned and checked for accuracy.

As you can see from just the small excerpt of a much larger data set presented in Table 16–1, most studies generate an enormous amount of information. The statistical actions experimental-type researchers use summarize this information and reduce the vast quantity of information to either general categories, summated scores, or single numeric indicators to represent the entire data set. Index development and descriptive statistical indicators such as the mean, mode, median, standard deviation, variance, and other statistically derived values reduce and summarize individual responses and scores. This summary process is often referred to as *data reduction* and is discussed in greater detail in Chapter 17. These scores are then submitted to other types of statistical manipulations to test hypotheses and make inferences and associations. As you will note in reading Chapter 17, quantitative approaches to the analysis of data are very well developed. Each statistical test has been developed over time and refined as computer technologies and the field of mathematics have expanded. There are clear and explicit rules as to when, how, and under what circumstances specific statistical analyses can be used. Because data must fit specific criteria for any one analysis technique to be used, you must be concerned with how you intend to measure concepts. The measurement process will determine the type of data you will obtain and, subsequently, the type of analyses you can perform.

ANALYSIS IN NATURALISTIC INQUIRY

If you were conducting a naturalistic inquiry, you would be analyzing phenomena as you observe them. As we discussed in Chapter 15, data collection and data analysis occur simultaneously and are integrated, each informing the other. In Chapter 18, we explore in more depth the specific analytic thinking of the qualitative researcher as it occurs in data collection and analysis and after completion of field work.

Qualitative analytic strategies during the conduct of research and at its conclusion transform the volumes of interview transcripts, audiotapes, videotapes, fieldnotes, and other observational information into meaningful categories, taxonomies, or themes to explain the meaning and underlying patterns of the phenomenon of interest. Analysis is based on narrative data that have been transcribed from either audiotape or notes in the form of fieldnotes, observational notes, or the investigator's personal diary. Or analysis may be based on visual data from photographs and videotaping. Transcription is time-consuming. You can count on an average of 6 to 10 hours of typing for each hour of tape, depending on the clarity of the tape and the typist's familiarity with the task. Explicit directions need to be given to the typist regarding how to handle pauses, silences, laughter, repetitive phrasing such as "uh huh," and if there is more than one participant, how to identify each on the tape. The investigator also must check for accuracy of transcription and attempt to resolve all misspellings and missing sections because of poor taping. The investigator must immerse himself or herself in the transcriptions, reading them several times. In the initial stages of any type of fieldwork, it is suggested that all tapes be transcribed. In a large-scale project that might yield a tremendous quantity of narrative, however, the investigator can be selective and transcribe a sampling from the tapes obtained later in fieldwork.[1]

The following very short excerpt is taken from more than 100 pages of narrative derived from a transcription of a 2-hour audiotape of a focus group discussion.[2] The focus group involved six occupational therapists practicing in rehabilitation hospitals throughout Pennsylvania. In the focus group, therapists were asked to review a case story and then discuss how they would proceed with treatment that might involve the selection and training in assistive devices for an older man after a cerebrovascular accident.

> MS. K: *I personally would probably start off trying to figure out where his left upper extremity was and its recovery and seeing if I could work with that to get it more functional first vs. whether I would go ahead right away and start issuing the suction things that would stabilize so that he could use his right hand.*
>
> MR. S: *The things that struck me are that, one, he had a stroke. So I started going through a list of equipment that he might use. But then a 67-year-old Italian gentleman from South Philadelphia starts to put other things in my mind. The fact that he lives alone made an impression, and that he had been a butcher. That starts to generate ideas for me.*
>
> MS. V: *You can sort of generally think for the most part that he was within functional limits in other perceptual cognitive areas. I guess from looking at the diagnosis end, that he is a right CVA and he does have a neglect. I would want to really assess his perceptual cognitive deficits a little further, especially in light of a nonfunctional left upper extremity, at that. So looking at the component end, I would want to further examine those areas.*
>
> MS. P: *Sensation! Also because it does not mention [in the vignette] anything about any sensation at all. It would be extremely important with neglect.*

This excerpt represents a precoded but corrected version of the transcript and illustrates the nature of "data" for the researcher in naturalistic inquiry. Combine this transcript with personal field notes that describe how the investigator is proceeding in data collection and analysis and with the recorded feelings and interpretations associated with each step. Think about all these written words. You quickly obtain an appreciation of the volume of information that is generated in this type of research and the need for complete emersion in the materials by the investigator in order to use each written datum as part of the analysis. These written forms of data quickly become voluminous, and the researcher must set up multiple files that are catalogued to facilitate cross-referencing.

Keeping materials organized is critical to facilitate preliminary analysis and testing of emergent ideas and to assist in the more formal write up of the report, ethnography, life history, or final manuscript. Researchers may use a variety of organizational schemes, depending on their own personal style, preferences, and scope and needs of the research. Some use computer word processing programs or computer-assisted coding programs such as Zyindex or Ethnograph.

In experimental-type research, computer-based statistical programs are routinely used to analyze data. In naturalistic inquiry, computer-based coding programs assist the investigator in *organizing, sorting,* and *manipulating* the arrangement of materials to facilitate the analysis. This is an important distinction. Nothing can replace the investigator's intense involvement with the information in qualitative analysis. The value of word processing programs and specially designed qualitative software packages is the ability to catalog, store, and rearrange large sets of information in different sorted files.

Even when computer assistance is sought, many investigators also depend on index cards or loose-leaf notebooks with tabs to separate and identify major topics and emerging themes. Some investigators establish multiple copies of narratives and cut and paste materials to organize and reorganize each written segment into meaningful segments.

Researchers need to maintain a running index section in the form of cards or in notebook form to help in cross-referencing and locating major passages and themes, their assigned codes, and their location in a transcript.

An index or codebook system can be developed in many different ways. An excerpt from one way of developing a codebook is provided in Table 16–3.

In this example, a code, its definition, and line location in the transcript was recorded. This particular excerpt was modified to relate to the example of a coded transcript in Table 16–4. The codebook reflects three broad analytic categories (medical, individual, and situational) that emerged as the major considerations by therapists when assistive devices were prescribed for older adults. In turn, each of these categories included a number of subcategories. Another analytic thread was the patterns of thinking used by therapists or their decision-making steps, and these were likewise analyzed and coded and subcategorized (only a small portion of which is displayed in Table 16–3). The listing of the location of each code in the transcript provides a quick guide to the text. The text can then be sorted by codes and smaller files can be created that reflect any one of the categories of interest.

As part of the information-gathering and storing process, the researcher may keep personal notes and diary-like comments and insights that provide a context from which to view and understand field notes at each stage of their engagement

TABLE 16-3.

Excerpt From Codebook for Focus Group Transcripts (as It Relates to Table 16–4)

Code	Definition	Line Location
Med	Medical factors	1282
(DIAG)	1. Diagnosis	
(FUNCT)	2. Function	
(HS)	3. Health status	
(PERC)	4. Perceptual	
INDV	Individual factors	
	1. Background characteristics	
(AGE)	a. Age	1284
(GEND)	b. Gender	
(SES)	c. Socioeconomic status	
(ETHN)	d. Ethnicity	1284
(REL)	e. Religion	
(ROLE)	2. Role	
	a. Past	1287
	b. Present	
	c. Future	
SITUAT		
(LIV)	1. Living situation	1286
(COM)	2. Community	1285
DEC-MAK	Decision-making process	
	1. Reference of order to thoughts	1281–1282
	2. Reflection	1288

in the field. In naturalistic inquiry, the investigator is an integral part of the entire research process. It is through the investigator and his or her interaction in the field and with informants that knowledge of the field emerges and develops. Thus, these personal comments are critical to the investigator's self-reflections that occur as part of the analytic process. As the investigator reflects on his or her own feelings, moods, and attitudes at each juncture of data collection and recording, he or she can begin to understand the lens through which he or she interpreted the cultural scene or slice of behavior at that point in time in the fieldwork. Organization of these notes and of the emergent interpretations is an important technique. The investigator can develop questions to guide further data collection and to explain observations as they occur. Summarizing or "memoing"[3] as one proceeds in the analysis also produces what has been called an "audit trail."[4] An audit trail indicates the key turning points in an inquiry in which the researcher has uncovered and revealed new understandings or meanings to the phenomena of interest. This trail can then be reviewed by others as a way of determining the credibility of the investigator's interpretations.

Researchers in naturalistic inquiry may use diverse approaches to analysis, and each labels his or her analytic activities quite differently. Think about the diverse philosophical traditions that compose this research tradition. It certainly

TABLE 16–4.

Excerpt From Focus Group Transcript With Preliminary Analytic Codes

Line	Codes MED: DIAG INDV: AGE/GEND/ETHN/ROLE SITUAT: LIV/COM DEC-MAK:
1281	The things that struck me are that, one,
1282	he had a stroke. So I started going through
1283	a list of equipment that he might use. But then
1284	a 67-year-old Italian gentleman from
1285	South Philadelphia starts to put other things in
1286	my mind. The fact that he lives alone made an
1287	impression, and that he had been a butcher. That
1288	starts to generate ideas for me.

is in keeping with these traditions that researchers would pursue multiple ways of examining narrative, given the varied ways of thinking and the different types of natural settings and materials qualitative researchers analyze.

Most researchers along the naturalistic continuum agree that it is reasonable and important for qualitative researchers to identify how they work and to describe their analytic process so that others can either agree or refute the final interpretation that emerges. Identifying the thinking and action processes of the analytic task in naturalistic inquiry increases the likelihood of producing accurate and more credible interpretations and understandings of the phenomena. In Chapter 18, we discuss the different analytic processes and techniques investigators use to increase the credibility of their interpretations.

REFERENCES

1. Agar MH: *The professional stranger: an informal introduction to ethnography,* New York, 1980, Academic Press.
2. Gitlin LN, Burgh D, Durham D: Factors influencing therapists' selection of adaptive technology for older adults in rehabilitation, *Legacies and lifestyles for mature adults,* Philadelphia, 1992, Temple University.
3. Glaser B, Strauss A: *The discovery of grounded theory: strategies for qualitative research,* Chicago, 1967, Aldine.
4. Guba EG: Criteria for assessing the trustworthiness of naturalistic inquiry, *Educ Commun Technol J* 29:75–92, 1981.

17

Statistical Analysis for Experimental-Type Research

- **What Is Statistical Analysis?**
- **Level 1: Descriptive Statistics**
- **Level 2: Drawing Inferences**
- **Level 3: Associations and Relationships**
- **Computer Statistical Programs**
- **Exercises**

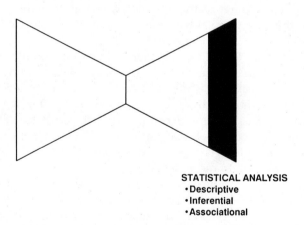

STATISTICAL ANALYSIS
- Descriptive
- Inferential
- Associational

Many newcomers to the research process often equate statistics with research. Hopefully you have come to understand that conducting statistical analysis is just one action process in research. Also by now you can see that there are no surprises in the research process along the experimental-type continuum. That is, the level of knowledge, the research problem, design, and procedures each logically lead to the selection of appropriate statistical actions. As this chapter demonstrates, the selection of a statistical procedure is determined by five

237

factors: the specific question you ask of your data set, the level of measurement of your data collection instrument, the quality of the information collected, the sampling procedures used, and the sample size obtained.

WHAT IS STATISTICAL ANALYSIS?

Statistical analysis is concerned with the organization and interpretation of data according to well-defined, systematic, mathematical procedures and rules. The term "data" refers to responses obtained through data collection to research questions such as how much, how many, how long, and how fast. In statistical analysis, data are represented by numbers. The value of numerical representation lies, in large part, in the clarity of numbers. This is a property that cannot always be exhibited by words. For example, let us say you visit your physician and he or she indicates that you need a surgical procedure. If the physician says to you that most patients survive the operation, you would want to know what is meant by "most." Does it mean 58 of 100 patients survive the operation? Eighty of 100? Numeric data provide a precise language to describe phenomena. As tools, statistical analyses provide a method to systematically analyze and draw conclusions to tell a story. Statistical analysis can be thought of as a stepping stone used by the experimental-type researcher to cross a stream from one bank (the question) to the other (the answer).

There are hundreds of different statistical procedures that can be used to analyze data. The primary objective of this chapter, however, is to familiarize you with the logic of choosing a statistical approach and the three levels of statistical analyses. Our focus is to provide an understanding of the logic behind the selection of procedures rather than of any one particular statistical test. We refer you to user-friendly books by Kachigan[1] and Huck et al.[2] in particular and the reference list for other texts that discuss the range of statistical analyses available and how to choose one in more detail.

Statistical analysis can be categorized into three levels: *descriptive, inferential,* or *associational.* The first level of statistical analysis, descriptive,[3] is used to reduce large sets of observations into more compact and interpretable forms. The second level of statistics is inference making, or the use of inferential statistics to draw conclusions about population parameters based on findings from a sample.[4] The third level of statistical analysis, associational, refers to a set of procedures designed to identify relationships between multiple variables and to determine whether knowledge of one set of data allows us to infer or predict the characteristics of another set of data. The purpose of these multivariate types of statistical analyses is to make causal statements and predictions. In Table 17–1, we have summarized the primary statistical procedures associated with each level of analysis. Each level of statistics corresponds to the particular level of knowledge available about the topic of interest and the specific type of question

TABLE 17-1.

Summary of Primary Tools of Each Level of Statistical Analysis

Level	Purpose	Selected Primary Statistical Tools
Descriptive statistics	Data reduction	Central tendency Mean Mode Median Variance Contingency tables Correlational
Inferential statistics	Inference to larger population	Parametric statistics Nonparametric statistics
Associational statistics	Causality	Multivariate analysis Multiple regression Discriminant analysis Path analysis

asked by the researcher. In Table 17–2, we have summarized the relationship of level of knowledge, question type, and statistical level of analysis.

Let us examine the purpose and logic of each level of statistical analysis in more detail.

LEVEL 1: DESCRIPTIVE STATISTICS

Think about all the numbers that are generated in a research study such as a survey. Each number provides information about an individual phenomenon but does not provide an understanding of a group of individuals as a whole. Remember our discussion in Chapter 14 on the four levels of measurement (nominal, ordinal, interval, and ratio)? Large masses of unorganized numbers, regardless of the level, are really not comprehensible and cannot in themselves

TABLE 17-2.

Relationship of Level of Knowledge to Question Type and Level of Statistical Analysis

Level of Knowledge	Question Type	Level of Statistical Analysis
Little to nothing known	Descriptive Exploratory	Descriptive
Descriptive information known	Explanatory	Descriptive
Little to nothing known as to relationships		Inferential
Relationships known	Predictive	Descriptive
Well-defined theory that needs to be tested	Hypothesis testing	Inferential Associational

answer your research question. Descriptive statistical analyses provide techniques to reduce large sets of data into smaller sets without sacrificing critical information. The data are more comprehensible if they are summarized into a more compact and interpretable form. This process is referred to as *data reduction* and involves the summary of data and their reduction to singular numeric scores. These smaller numeric sets are then used to describe the original observations. A descriptive analysis is the first action a researcher takes to understand the data that have been collected. There are several techniques for reducing data, referred to as descriptive statistics. Included in this category are statistics such as *frequencies, central tendencies (mean, mode,* and *median), variances, contingency tables,* and *correlational analyses.* These statistics are concerned with direct measures of population characteristics. For example, suppose you are in the process of developing a new drug abuse prevention program in your local school district. Without knowing the age range and level of knowledge related to drug abuse held by students, you would be at a loss in your planning effort. You therefore might undertake a simple descriptive research project to obtain the ages and level of knowledge held by the students. However, ages and scores from each individual will not be useful to you if you have 200 youth. Needing a better way to work with your data, you would then conduct a descriptive analysis of the age ranges and of the average knowledge scores. You might arrange ages according to how many youth fit into each age category (such as 10 to 12, 13 to 15) and so forth. This *frequency analysis* tells you the number of students in each group and would be more valuable to you than knowing each age or the average age in your group.

Let us look at each type of descriptive statistic.

Frequency Distribution

The first and most basic descriptive statistic is the frequency distribution. This term, illustrated earlier in our example of organizing student ages, refers to the distribution of values for a given variable and the number of times each value occurs. The distribution reflects a simple tally or count of how frequently each value of the variable occurs among the set of measured objects. Frequency distributions are usually arranged so that the values of the variable range from lowest to highest or highest to lowest. Let us return to our previous example. The following is an illustration of frequency distribution of student ages. As you can see, 100 students fall between the ages of 10 and 12, 50 fall between the ages of 13 and 15, and 50 fall between the ages of 16 and 18.

Frequency Distribution of Ages of Students (N = 200)		
10–12	**13–15**	**16–18**
100	50	50

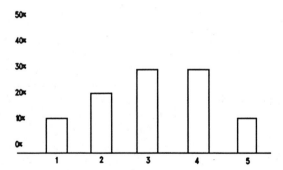

FIG 17-1.
Histogram depicts distribution of scores by percentage of responses in each category.

For variables in which there is a very large number of categories, it is often more efficient to group data based on a theoretical basis. Frequencies can be represented in two ways: histograms or bar graphs and polygons (dots connected by lines). Frequencies provide information regarding two basic aspects of the data that have been collected. They allow the researcher to identify the most frequently occurring class of scores and any pattern in the distribution of those scores. In addition to a count or tally, frequencies also can produce what are referred to as "relative frequencies." These are the observed frequencies converted into percentages based on the total number of observations. For example, relative frequencies tell us at a glance what percentage of subjects had a score on any given value.

Let us examine a hypothetical data set to illustrate. You have just measured the life satisfaction of a large group of persons (N = 1,000) over the age of 65 years. Examining raw scores individually would not provide an understanding of patterns or how the group behaved on this variable. You first decide to

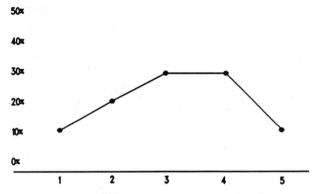

FIG 17-2.
Polygon depicts distribution of scores by percentage of responses in each category.

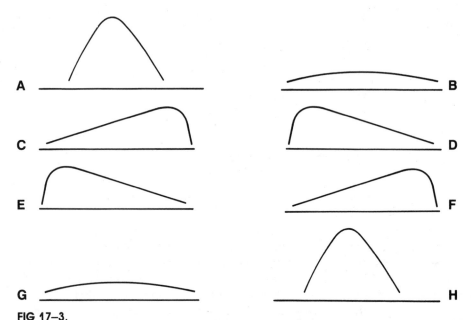

FIG 17–3.
A and **B,** symmetric distribution. **C** and **D,** nonsymmetric distribution. **E,** positively skewed distribution. **F,** negatively skewed distribution. **G,** platykurtosis (flatness). **H,** lepokurtosis (peakedness).

represent your findings visually to obtain an overview of the nature of the data you have collected. Based on self-report, you have rated each respondent as belonging to one of five categories of life satisfaction represented by the numbers in parentheses: very satisfied (5), satisfied (4), neutral (3), unsatisfied (2), and very unsatisfied (1). Figure 17–1 depicts a histogram of the distribution of the scores by percentage of responses in each category (relative frequency).

A polygon (Fig 17–2) can be drawn to present the same information. A dot is plotted for the percentage value, and lines are drawn among the dots to yield a picture of the shape of the distribution, as well as the frequency of responses in each category.

Frequencies can be described by the nature of their distribution. They can be either symmetric, in which both halves of the distribution are identical (Fig 17–3, A and B), or nonsymmetric (Fig 17–3, C and D), skewed to the right (Fig 17–3, E), or skewed to the left (Fig 17–3, F), or kurtosis, which is a frequency distribution characterized by flatness (platykurtic; Fig 17–3, G) or peakedness (leptokurtic; Fig 17–3, H).

Measures of Central Tendency

A frequency distribution reduces a large collection of data into a relatively compact form. Although we can use terms to describe a frequency distribution as bell-shaped, kurtosis, or skewed to the right, we can also summarize the frequency distribution itself using specific numerical values. These values, or measures of central tendency, provide us with information regarding the most typical or representative scores in a group. The three basic measures of central tendency are:

- Mode
- Median
- Mean.

Mode

In most data distributions, observations tend to cluster heavily around certain values. One logical measure of central tendency is the value that occurs most frequently. This value is referred to as the modal value, or the mode. For example, consider the data collection of nine observations such as the following values:

$$9 \quad 12 \quad 15 \quad 15 \quad 15 \quad 16 \quad 16 \quad 20 \quad 26$$

In this distribution of scores, the modal value is 15 because it occurs more than any other score. The mode, therefore, represents that value of the variable that occurs most often and does not refer to the frequency associated with that value.

In a distribution based on data that have been grouped into intervals, the mode is often taken as the midpoint of the interval that contains the highest frequency of observations. For example, refer back to the frequency distribution of student ages (p 240). The age interval in which the most cases are observed is the 10- to 12-year range. Because the exact ages of the individuals within that category are not known, we select 11 as our mode, because it is the midway point in the category with the highest frequency.

Some distributions can be characterized as *bimodal* in that two values occur with the same frequency. Let us use the age distribution of our students once again. This time let us change the values to represent the following age distribution:

Age (yr)	Frequency
10–12	50
13–15	75
16–18	75

In this distribution, two categories have the same frequency, and you should also note that both categories have the highest frequencies. Therefore, this is a bimodal distribution, with 14 being the value of one mode and 17 being the value of the second mode.

The advantage of the mode is that it can easily be obtained as a single indicator of a large distribution. Also, the mode can be used for statistical procedures with categorical variables (numbers assigned to a category, such as male = 1, female = 2).

Median

The second measure of central tendency is the median, which refers to that point on a scale at which 50% of the cases fall above and below. The median lies at the midpoint of the distribution. The median is determined by first arranging a set of observations from lowest to highest in value. Then the middle value is singled out so that 50% of the observations fall above and below that value. Consider the following values:

$$22\ 24\ 24\ 25\ 27\ 30\ 31\ 35\ 40$$

Here the median is 27 because half the scores fall below the number 27 and half are above. In an odd number of values, as in the above case of nine, the median is one of the values in the distribution. When an even number of values occurs in a distribution, the median may or may not be one of the actual values because there is no middle number. In other words, an even number of values exist on both sides of the median. Consider the following values:

$$22\ 24\ 24\ 25\ 27\ 30\ 31\ 35\ 40\ 47$$

Here the median lies between the fifth and sixth values. The median is therefore calculated as an average of the scores surrounding it. In this case, the median is 28.5 because it lies between the values of 27 and 30. If the sixth value had been 27, the median would have been 27.

In the case of a frequency distribution based on grouped data, the median can be reported as that interval in which the cumulative frequency equals 50% (or as a midpoint of that interval).

The major advantage of the median is that it is insensitive to extreme scores in a distribution. That is, if the highest score in the set of numbers just shown had been 85 instead of 47, the median would not be affected. Income is a good example of how the median is a good indicator of central tendency because it is not effected by extreme values.

Mean

The mean, as a measure of central tendency, is the most fundamental concept in statistical analysis. The mean serves two purposes. First, it serves as a data reduction technique in that it provides a summary value for an entire distribution. Second, and most important, it serves as a building block for many other statistical techniques. As such, the mean is the most important measure of central tendency. There are many common symbols for the mean. Among them are:

$$Mx = \text{mean of variable x}$$
$$My = \text{mean of variable y}$$
$$\bar{X} = \text{x bar or mean value of variable x}$$
$$M = \text{mu, mean of a sample}$$

The formula for calculating the mean is simple:

$$M = \frac{EX_i}{N}$$

where EX_i = sum of all values, M = mean, and N = total number of observations.

The major advantage of the mean over the mode and median is that in calculating this statistic, the numerical value of every single observation in the data distribution is taken into account and used. In other words, when the mean is calculated, all values are summed and then divided by the number of values. However, this very strength can be a draw back with highly skewed data in which there are "outliers," or extreme scores.

Let us consider an example to illustrate this point. Suppose you have just completed teaching a continuing education course in cardiopulmonary resuscitation (CPR). You then test your students to determine their competence in CPR knowledge and skill. Of a possible 100, the following scores were obtained:

$$100$$
$$100$$
$$100$$
$$95$$
$$95$$

To calculate the mean, you would add each value, then divide by the total number of values ($100 + 100 + 100 + 95 + 95 = 490/5 = 98$). The mean score of your group is 98. You are satisfied with the scores and with the high level of knowledge and skill. Now let us see what happens in the following distribution of test scores:

$$100$$
$$100$$
$$100$$
$$95$$
$$95$$
$$30$$

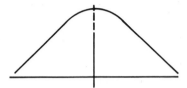

FIG 17–4.
Normal curve where mean, median, and mode are in same location.

The mean would then be calculated as $100 + 100 + 100 + 95 + 95 + 35 = 425/6 = 70.8$. The mean (70.8) presents quite a different picture even though only one member of the group scored poorly. If only the mean were reported, you would have no way of knowing that the majority of your class did well and that one individual or outlier score was responsible for the lower mean.

So, now that you are aware of three measures of central tendency, how do you determine which measure or measures you should calculate? Although investigators often calculate all three measures, all might not be useful. Their usefulness depends on the purpose of the analysis and the nature of the distribution of scores. In a normal curve, the mean, median, and mode are in the same location (Fig 17–4). In this case, it is most efficient to use the mean, because it is the most widely used measure of central tendency and forms the foundation for subsequent statistical calculations.

However, in a skewed distribution, the three measures of central tendency fall in different places, as shown in Figure 17–5. In this case, you need to examine all three measures and select the one that most reasonably answers your question without providing a misleading picture of the findings.

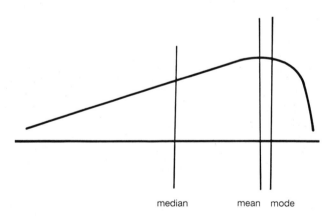

median   mean   mode

FIG 17–5.
Skewed distribution where the three measures of central tendency fall in different places.

Measures of Variability

We have learned that a single numerical index such as the mean, median, or mode can be used to describe a large frequency of scores. However, each has certain limitations especially in the case of a distribution with extreme scores.

Most groups of scores on a scale or index differ from one another or have what is termed variability. Variability is another way of summarizing and characterizing large data sets. The measure of variability simply indicates the degree of dispersion among a set of scores. If scores are very similar, there are little dispersion and little variability. On the other hand, if scores are very dissimilar, there is a high degree of dispersion. Even though a measure of central tendency provides a numerical index of the average score in a group, it is also helpful to know how the scores vary or are dispersed around the measure of central tendency. Measures of variability provide additional information about the scoring patterns of the entire group.

For example, let us consider the following two groups of IQ scores:

Group 1	Group 2
102	128
99	78
103	93
96	101

In both, the mean IQ score is equal to 100. However, the variability or dispersion of scores around the mean is quite different. The scores in group 1 are more homogeneous, whereas in the second group scores are more variable, dispersed, or heterogeneous (Fig 17–6). Therefore, just knowing the mean would not help to ascertain differences in the two groups even though they existed.

To more fully describe a data distribution, a summary measure of the variation or dispersion of the observed values is needed. There are two basic measures of variability, the *range* and the *standard deviation.*

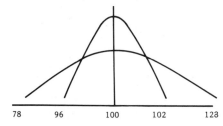

FIG 17–6.
Homogeneous and heterogeneous IQ scores.

Range

The range represents the simplest measure of variation. It refers to the difference between the highest and lowest observed value in a collection of data. The range is actually a crude measure in that it does not take into account all values of a distribution, only the lowest and highest. In other words, it does not indicate anything about the values that lie between these two extremes. In the group 1 data listed earlier, the range is 96 to 103, or 7. In the group 2 data, the range is 78 to 128, or 50.

Standard Deviation

The standard deviation is the most widely used measure of variability. It is an indicator of the average deviation of scores around the mean. In reporting the standard deviation, researchers often use SD, s, sigma, or σ. Like the mean, it is calculated by taking into consideration every score in a distribution. The standard deviation is based on distances of sample scores from the mean score and equals the square root of the mean of the squared deviations. It represents the sample estimate of the population standard deviation and is calculated by the following formula:

$$S = \sqrt{\frac{\epsilon(y - y)^2}{n - 1}}$$

Look at the following calculation of the standard deviation for these observations and a mean of 15.5 (or 16):

$$14 \ 21 \ 15 \ 12$$

$$S = \sqrt{\frac{(14 - 16) + (21 - 16) + (15 - 16) + (12 - 16)}{4 - 1}}$$

$$= \sqrt{\frac{4 + 25 + 1 + 16}{4 - 1}}$$

$$= \sqrt{\frac{46}{3}}$$

$$= 3.9$$

The first step is to compute deviation scores for each subject. Deviation scores refer to the difference between an individual score and the mean. Then each deviation score is squared. If the deviation scores were simply added without being squared, the sum would equal zero. This is the case because the deviations above the mean always balance exactly those deviations below the mean. The standard deviation overcomes this problem by squaring each deviation score before summary. Third, the squared deviations are summed, the result is divided by one less than the number of cases, and then the square root is obtained. The square root takes the index back to the original units. In other words, the

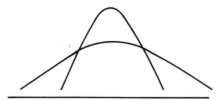

FIG 17–7.
Standard deviation, an index of variability of scores in a data set, tells how much scores deviate, on average, from the mean.

standard deviation is in the units that are being measured.

The standard deviation is an index of variability of scores in a data set. It tells us how much scores deviate on the average from the mean.

For example, if two distributions have a mean of 25, but one has an SD of 70 and the other of 30, we know the second sample is more homogeneous. That is, we know that the scores cluster more closely around the mean or that there is less dispersion. Figure 17–7 visually illustrates this point.

Bivariate Descriptive Statistics

Thus far we have described descriptive data reduction approaches for one variable. This class of procedures is referred to as a univariate approach. Another aspect of describing data involves looking for relationships among two or more variables. We describe two methods here: contingency tables and correlation.

Contingency Tables

One method for describing a relationship between two variables, or a bivariate relationship, is a contingency table, which is also referred to as cross tabulation.

A contingency table is a two-dimensional frequency distribution that is used primarily with categorical data. Let us return to our example of measuring life satisfaction in the elderly. Suppose you were interested in life satisfaction and whether age has any relationship to this concept. In other words, you are investigating the relationship between two variables: age and life satisfaction. To organize your age variable, you specify four age intervals: 65 to 70, 71 to 75, 76 to 80, and 81 and older. You then develop the frequency distribution:

Frequency Distribution (N = 1,000)	
≥81	200
76–80	200
71–75	200
65–70	200

Your frequency analysis demonstrates an equal age distribution among your respondent group. You then construct a contingency table to ascertain how many respondents in each age interval scored in each category of life satisfaction:

	1	2	3	4	5
≥81	40	40	40	40	40
76–80	40	40	40	40	40
71–75	40	40	40	40	40
65–70	40	40	40	40	40

Although highly unlikely, this contingency table tells you that there was an equal distribution of scores in each life satisfaction category for each age group. This indicates no relationship between age and life satisfaction. There are a number of statistical procedures that researchers use to determine the nature of the relationships displayed in a contingency table.

Correlational Analysis

A second method of determining relationships among variables is correlational analysis. This approach examines the extent to which two variables are related to each other. There are numerous correlational statistics. Selection is dependent primarily on level of measurement and sample size. In a correlational statistic, an index is calculated that describes the *direction* and *magnitude* of a relationship. Three types of directional relationships can exist among variables: positive correlation, negative correlation, and no correlation, termed zero correlation. Positive correlation indicates that as the value of one variable increases or decreases, so does the other. Conversely, negative correlation indicates that each variable is related in an opposing direction. As one increases, the other decreases. For example, the relationship between age and height in children younger than 12 years demonstrates a positive correlation, whereas the relationship between illiteracy and education is a negative correlation.

To indicate magnitude or strength of a relationship, the value that is calculated in correlational statistics ranges from -1 to $+1$. The value of -1 indicates a perfect negative correlation, and $+1$ signifies a perfect positive correlation. By perfect, we mean that each variable changes at the same rate as the other. For example, suppose you found a perfect positive correlation $(+1)$ between years of education and reading level. This statistic would tell you that for each year (unit) of education, reading level improved 1 unit. A zero correlation is demonstrated in the previous table, which displays the relationship between life satisfaction and age. As you can surmise, the closer a correlational statistical value falls to $|1|$ (absolute value of 1), the stronger the relationship (Fig 17–8). As the value approaches zero, the relationship weakens.

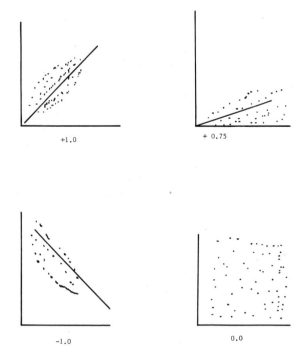

FIG 17–8.
Correlational analyses.

Two correlation statistics are used frequently in social science literature: the *Pearson product-moment correlation* and *Spearman rho*. Both statistics yield a value between -1 and $+1$, but the Pearson r is calculated on interval level data, whereas the Spearman rho is used with ordinal data.

To illustrate the use of the Pearson r, let us examine a study by Rozier et al.[5] investigating job satisfaction in occupational therapy faculty. The researchers examined numerous variables measured with interval level data to determine which were related to job satisfaction in a sample of 538 faculty. To ascertain the extent of the relationships between selected demographic variables and attitudes, the researchers conducted a series of Pearson r calculations. They found that the strongest correlation (r = 0.38) existed between the highest degree earned and the attitude "Research is exciting." The weakest correlation (r = $-.014$) existed between previous opportunity to teach in the clinical setting and the attitude "Seeing students grow is exciting." Note that the first correlation is positive and the second is negative. If the investigators had used ordinal data in their measurements, they would have used the Spearman rho to examine these relationships.

You may be asking a question about whether the findings are valuable. Could the relationships reported by Rozier et al.[5] be a function of chance? Are these important in light of such seemingly low correlations? To answer these questions, investigators first submit the data to a test of significance. A test of significance determines the extent to which a finding occurs by chance. The investigator selects a level of significance, a statement of the expected degree of accuracy of the findings, based on the sample size and on convention in the literature and then looks up the correlation coefficient in a table in the back of a statistics book. If the correlation value exceeds the number listed in the table, the investigator can assume with a degree of certainty that the relationship was not caused by chance.

Let us consider how this is done. Suppose that Rozier et al.[5] selected 0.05 as their confidence level. This number means that 5 times out of 100, the results will be caused by chance. They then go to a table of critical values and find that their number of 0.38 exceeds the value that they need to determine that the finding is significant or, in other words, was not caused by chance. They therefore conclude that 0.38 is significant at the 0.05 level and report it to the reader as such.

There are other statistical tests of association that are used with different levels of measurement besides the two we have discussed here. For example, investigators often calculate the point biserial statistic to examine a relationship between a nominal variable and an interval level measure. When two nominal variables are calculated, the phi correlation statistic is often selected by researchers as an appropriate technique.[6]

LEVEL 2: DRAWING INFERENCES

Although descriptive statistics are useful for summarizing univariate and bivariate sets of data, researchers also want to determine the extent to which observations of the sample are representative of the larger population from which the sample was selected. Inferential statistics provide the action processes for drawing conclusions about a population given the data obtained from a sample. Remember that the purpose of testing or measuring a sample is to gather data that allow us to make statements about the characteristics of a population (Fig 17–9). Statistical inference, which is based on probability theory, is the

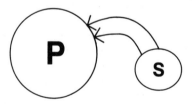

FIG 17–9.
Relationship between sample *(S)* and population *(P).*

process of generalizing from samples to whole populations. The tools of statistics help identify valid generalizations and those that are likely to stand up under further study. Thus, the second major role of statistical analysis is to make inferences. Inferential statistics include statistical techniques for evaluating differences between sets of data. These techniques are used to evaluate the degree of precision and confidence of one's measurements.

Let us consider an example to illustrate how inferential statistics help to determine the extent to which findings about a sample apply to the population from which it was selected. Suppose you are a mental health provider and are interested in testing the effect of a new rehabilitation technique on the functional level of persons with schizophrenia who are hospitalized in long-term care institutions. Not only do you want to evaluate the technique for its effectiveness on one unique group, but you think that it has universal merit and have decided to determine its value to the population of persons with schizophrenia residing in institutional settings. As we discussed in Chapter 12, your first order of business is to define your population with both inclusion and exclusion criteria. You then randomly select a sample from your population through the appropriate random sampling techniques and randomly assign the sample to experimental and control conditions. You pretest all subjects, expose the experimental group to the intervention, and then post-test all subjects. Hypothetically your measure of functional status ranges from 1 to 10, with 1 being least functional and 10 being most functional. Now you have a series of scores: pretest scores from both groups and post-test scores from both groups.

Let us assign hypothetical means to each group for illustration. The experimental group has a pretest mean of 2 and a post-test mean of 6, whereas the control group has a pretest mean of 3 and a post-test mean of 5. Of course, in a real study, you would calculate measures of dispersion as well. However, for instructive purposes, we will omit that measure.

To answer your research question, you must examine these scores at several levels. First, you want to know if the groups are equivalent before the intervention. You hypothesize that there would be no difference between the two groups before participation in the intervention. Also, you would hypothesize that the groups adequately represent the population. Visual inspection of your pretest means shows different scores, but is this difference occurring by chance or as a result of actual group differences? To determine whether the difference in mean scores is caused by chance, you would select a statistical analysis (for example, an independent t-test) to compare the two sets of pretest data (experimental and control group mean scores). At this point you hope to find that the scores are equivalent and therefore that both sample groups represent the population.

Second, you want to know whether the change in scores at the post-test are a function of chance or what is anticipated to occur in the population. You therefore choose a statistical analysis to evaluate these changes. (The analysis of covariance [ANCOVA] is the statistic of choice for the true experimental design). In actuality, you are hoping that the experimental group demonstrates change on the dependent or outcome variable at post-test time. You hypothesize that the change will be greater than that which may occur in the control group and greater than the scores observed at pretest time. Thus, as a consequence of participating in the experimental group condition, you want to show that the sample is significantly different than the control group and thus the population from which it was drawn. In actuality you are testing two phenomena: *inference,* the extent to which the samples reflect the population both at pretest and post-test time, and *significance,* the extent to which group differences are a function of chance.

If you recall from Chapter 12 on sampling, the population refers to all possible members of a group defined by the researcher. A sample refers to a subset of a population from which the researcher wants to be able to generalize. The accuracy of inferences from sample to population depends on how representative the sample is of the population. The best way to ensure representativeness is to use probability sampling techniques.

To use an inferential statistic, you must follow five action processes, introduced in Chapter 12. Let us review them here:

- Action 1: State the hypothesis
- Action 2: Select a significance level
- Action 3: Compute a calculated value
- Action 4: Obtain a critical value
- Action 5: Reject or fail to reject the null hypothesis

Action 1: Stating the Hypothesis

Stating a hypothesis is both a simple and a complex action process. In experimental-type designs that test differences between the population and the sample, most researchers state a hunch of what they expect to occur. This statement is called a *working hypothesis.* However, for statistical analysis, a working hypothesis is transformed into the *null hypothesis.* The null hypothesis is a statement of no difference between or among groups. In some studies, the investigator hopes to accept the null hypothesis, whereas in other studies, the hope is to reject it. Initially, as in the example of the experimental design given earlier, we hope to accept the null hypothesis of no difference between the pretest mean scores of the experimental and control groups. That is, we do not want our sample scores to differ from the population scores and for the experimental and control groups to differ from one another at baseline or before the introduction of the treatment or intervention. We pose and test the null hypothesis and hope that in testing it, our statistical value will fall below the critical value in the statistical tables. However, for our post-test and change scores, we hope to reject the null hypothesis. We want to find differences between the post-test scores and a change in the experimental group scores that are not representative of the population that has not received the intervention.

Why is the null hypothesis used? Theoretically, it is impossible to "prove" a relationship among two or more variables.[6] It is only possible to negate the null hypothesis of no difference.[7, 8] Nonsupport for the null hypothesis is similar to stating a double negative. If it is not "not raining," it follows logically that it is raining. Applied to research, if there is no "no difference among groups," differences among groups can be assumed, although not proved.

Action 2: Selecting Significance Level

Level of significance defines how rare or unlikely the sample data must be before the researcher can reject the null hypothesis. That is, the level of significance indicates how confident you are that the findings regarding the sample are not caused by chance. For example, if you select a significance level of 0.05, you are 95% confident that your statistical findings did not occur by chance. If you were to repeatedly draw different samples from the same population, you would find similar scores 95 out of 100 times. Similarly, a confidence level of 0.1 indicates that the findings may be caused by chance in one out of every ten chances.

As you can see, the smaller the number, the more confidence the researcher has in his or her findings, and the more credible the results. Because of the nature of probability theory, you can never be sure that your findings are 100% accurate. Significance levels are selected by the researcher based on sample size, level of measurement, and conventional norms in the literature. As a rule of thumb, the larger the sample size, a smaller numerical value in the level of significance is used. This principle makes a lot of sense if you think about it. If you have a very small sample size, you risk obtaining a group that is not highly representative of the larger population. Thus, your confidence level drops. A very large sample size includes more elements from the population, and thus the chances of representation and confidence of findings increase. You therefore can use a very stringent level of significance (0.01 or smaller).

Because in statistical inference researchers are dealing with probabilities, two types of statistical inaccuracy, or error, can contribute to the inability to claim full confidence in findings. These are called *type I* and *type II errors*.

Type I Error

In a type I error (also called *alpha* error[4]), the researcher rejects the null hypothesis when it is true (in other words, claims a difference between groups when, if the entire population were measured, there would be no difference). This error can occur when the most extreme members of a population are selected by chance in a sample. Assume, for example, that you set the level of significance at 0.05, indicating that five times out of 100 the null hypothesis could be rejected when it is accurate. Because the probability of making a type I error is equal to the level of significance chosen by the investigator, reducing the level of significance will reduce the chances of making this type of error. Unfortunately, as the probability of making a type I error is reduced, the potential to make another type of error increases.

Type II Error

A type II (*beta*[4]) error occurs if the null hypothesis is accepted when it actually should be rejected. In other words, the researcher fails to ascertain group differences when, in fact, they occur. If you made a type II error, you would conclude, for example, that the intervention did not have a positive outcome on the dependent variable when it actually did. The probability of making a type II error is not as apparent as that of making a type I error.[2] It is based in large part on the power of the statistic to detect group differences.[4]

Type I and II errors are mutually exclusive. That is, you make either one or the other. Furthermore, it is difficult to determine if either error has been made, because actual population parameters are not known by the researcher. It is often considered more serious to make a type I error, because you are claiming a significant relationship or outcome when there is none. Because other researchers or practitioners may act on that finding, you want to be very careful about type I errors. On the other hand, not recognizing a positive effect from a treatment program, for example, can also have serious consequences for practice.

Action 3: Computing a Calculated Value

To test a hypothesis, one must choose and calculate a statistical formula. The selection of a statistic is based on your research question, level of measurements, the number of groups that you are comparing, and the sample size.

There are two classifications of inferential statistics from which an investigator chooses a statistic: *parametric* and *nonparametric procedures*. Both parametric and nonparametric procedures are similar in that they both (1) test hypotheses, (2) involve a level of significance, (3) require a calculated value, (4) compare against a critical value, and (5) conclude with decisions about the hypotheses.

Parametric Statistics

Parametric statistics are mathematical formulas that test hypotheses based on three assumptions:

1. The samples come from populations that are normally distributed. In other words, the scores form a normal curve when plotted on X and Y axes.

2. There is homogeneity of variance. In other words, variances within groups are the same. Homogeneity is displayed by scores in one group having about the same degree of variability as the scores in other groups.
3. The data generated from the measures are interval level.

Parametric statistics can test the extent to which numerous sample structures are reflected in the population. For example, some statistics test differences between two groups only, whereas others test differences among many groups. Some test *main effects* (the direct effect of one variable on another), whereas other tests have the capacity to test both main and *interactive effects* (the combined effects that several variables have on another variable). Some statistical action processes test group differences at one time only, whereas others test differences over time. We cannot present the full spectrum of parametric statistics in this book, so we will discuss the most frequently used statistical tests. Most researchers will attempt to use parametric tests when they can because they are the most robust of the inferential statistics. Robust tests are those that are most likely to detect a significant effect or increase power and decrease type II errors.

Nonparametric Statistical Procedures

Nonparametric statistical procedures are statistical formulas used to test hypotheses when:

1. Normality of variance in the population is not assumed.
2. Homogeneity of variance is not assumed.
3. The data generated from measures are ordinal or nominal.
4. Sample sizes may be small.

Most parametric statistics have nonparametric analogues. Therefore, nonparametric statistics can be used to test hypotheses from as wide a variety of designs, as can parametrics. Because they are less robust that parametric tests, researchers tend not to use them unless they believe that the assumptions necessary for the use of parametric statistics have been violated.

Deciding on a Statistical Test

Choosing a statistical test is based on several considerations:

1. What is the research question?
   a. Is it about differences?
   b. Is it about degrees of a relationship between variables?
   c. Is it an attempt to predict group membership?

2. How many variables are being tested, and what types of variables are they?
    a. How many variables do you have? For example, 1 independent variable and 1 dependent variable equals bivariate analysis. If there is more than 1 independent variable and more than 1 dependent variable, you would do a multivariate analysis.
3. What is the level of measurement?
    a. Interval level can be used with parametric procedures.
4. Are variables continuous or discrete?
5. What is the nature of the relationship being investigated between two or more variables?
6. How many groups are being compared?
7. What are the underlying assumptions about the distribution of a measurement in the population from which the sample was selected?
8. What is the sample size?

The answers to these questions guide the researcher to the selection of specific statistical procedures. A discussion of the tests frequently used by researchers follows. This section should be used only as a guide. We refer you to other texts and, in particular, the statistical decision trees constructed by Andrews et al.,[9] which should help lead you to the selection of appropriate statistical techniques.

Popular Parametric Statistical Procedures: Univariate Approaches

Three techniques are used to compare two or more groups to see whether the differences between group means are large enough to assume that the corresponding population means are different: *t-tests, analysis of variance,* and *multiple comparisons.*

The t-tests are the most basic statistical procedure in this grouping. They are used to compare *two* sample means. For example, suppose you want to compare the intelligence level of physical therapy students with occupational therapy students. You might begin by administering the Wechsler Adult Intelligence Scale[10] to a randomly selected sample of students. For reliability, you administer the test on two occasions and obtain two mean scores from both groups. From these scores, you would then calculate the mean scores for each group. The following depicts the hypothetical means of the groups:

Group 1 (PT)	Group 2 (OT)
108	106
114	132
$\bar{X} = 117.5$	$\bar{X} = 125.6$

At first glance it appears as if group 2 has the larger mean. However, it is not that much larger than the mean from group 1. Therefore, the statistical question becomes: Are the two sample means sufficiently different to allow the researcher to conclude, with a high degree of confidence, that the population means are different from one another (even though we will never see the actual population means)? The t-test provides an answer to that question. If the researcher finds a significant difference between the two sample means, the null hypothesis will be rejected. The t-test can be used only when the means of two groups are compared. For studies with more than two groups, one must select other statistical procedures. Remember also that t-tests, like all parametric statistics, must be calculated with interval level data and should be selected only if the researcher believes that the assumptions for the use of parametric statistics have not been violated. The t-test yields a t value that is reported as $t = x$, $p = 0.05$. where x is the calculated t value and p is the level of significance set by the researcher. There are several variations of the t-test; some test unidirectional comparisons, whereas others test bidirectional comparisons. Variations also occur in t-tests related to the timing of its use and other factors.

The one-way analysis of variance (ANOVA) serves the same purpose as the t-test: to compare sample group means to determine if a significant difference can be inferred in the population. However, ANOVA can handle two or more groups. That is, ANOVA is an extension of the t-test for a two or more group situation. The null hypothesis for an ANOVA, as in the t-test, states that there is no difference between the two or more population means. The procedure is also similar to the t-test. The original raw data are put into a formula to obtain a calculated value. The resulting calculated value is compared against the critical value, and the null hypothesis is rejected if the calculated value is larger than the tabled critical value or accepted if the calculated value is less than the critical value. Computing the one-way ANOVA yields an F value that may be reported as $F(a, b) = x$, $p = 0.05$, where x equals computed F value, *a* equals group degrees of freedom, *b* equals sample degrees of freedom, and *p* equals level of significance. Degrees of freedom refer to the "number of values, which are free to vary"[6] in a data set.

There are many variations of ANOVA, some of which test relationships when variables have multiple levels and some of which examine complex relationships among multiple levels of variables. For example, suppose you were interested in determining the effects of rehabilitation intervention, family support, and functional level on the recovery time from closed head injury. Measuring the relationship between each independent variable and the outcome would be valuable. However, it would seem prudent to consider the interactive effects of the variables on the outcome. Several variations of the ANOVA (such as analysis of covariance [ANCOVA] exist that should be considered.

When a one-way ANOVA is used to compare three or more groups, a significant F value means that the sample data indicate that the researcher should reject the null hypothesis. However, the F value, in itself, does not tell you which of the group means are significantly different, only that one or more are different. There are several procedures referred to as *multiple comparison* or *post hoc comparisons* to determine which group is greater than the others. These procedures are computed after the occurrence of a significant F value and are capable of identifying which group or groups differ among those being compared.

Each of the tests mentioned thus far has a nonparametric analogue. For example, the nonparametric analogue of the t-test for categorical data is the chi-square. For some of the nonparametric tests, the critical value may have to be larger than the computed statistical value for findings to be significant.

Action 4: Obtaining a Critical Value

The researcher must look up a critical value in an appropriate statistical table in the back of a statistical book. The critical value is a criterion related to the level of significance and tells you what number you must derive from your statistical formula to have a significant finding.

Action 5: Rejecting or Failing to Reject the Null Hypothesis

The final action process is the decision about whether to accept or reject the null hypothesis. So far we have indicated that a statistical formula is selected, along with a level of significance. The formula is calculated, yielding a numerical value. So how do researchers know whether to accept or reject the null hypothesis based on the obtained value? Before the use of computers, the researcher would set a significance level and calculate degrees of freedom for his or her sample or number of sample groups (or both). Degrees of freedom are closely related to sample size and number of groups. Calculating degrees of freedom is dependent on the statistical formula used. Suffice it to say that by examining the degrees of freedom in a study, you can closely ascertain sample size and number of comparison groups without reading anything else. Just as an example, in the t-test, degrees of freedom are calculated on the sample size only. Group degrees of freedom are not calculated because only two groups are analyzed, and, therefore, only one group mean is free to vary. When degrees of freedom for sample size are calculated, the number 1 is subtracted from the total sample ($DF = n - 1$), indicating that all measurement values with the exception of one are free to vary. For the F ratio, degrees of freedom are calculated on groups means as well because more than two groups may be compared. Once the researcher has calculated the degrees of freedom and the statistical value, he or she then goes to the table that illustrates the distribution for the statistical values that were conducted. The critical values are located by observing the value that is listed at the intersection of the calculated degrees of freedom and the level of significance. If the critical value is larger than the statistic calculated, in most cases the researcher accepts the null hypothesis (in other words, no significance differences between groups). If the calculated value is larger than the critical value, the researcher rejects the null hypothesis and within the confidence level selected accepts that the groups differ.

With the increasing use of computers, researchers rarely make use of tables any more. The computer itself is capable of calculating all values and further identifying the p value at which the calculated statistical value will be significant.

LEVEL 3: ASSOCIATIONS AND RELATIONSHIPS

The third major role of statistics is the identification of relationships between variables and whether knowledge about one set of data allows us to infer or predict characteristics about another set of data. Included among these statistical tests are factor analyses, discriminant function analysis, multiple regression, and modeling techniques. The commonalities among these tests is that they all seek to predict one or more outcomes from multiple variables. Some of the techniques can further identify time factors, interactive effects, and complex relationships among multiple independent and dependent variables.

To illustrate this level of statistical analysis, let us consider the study by Palmore[11] in which he investigated predictors of outcome in nursing homes. Palmore obtained data on 22 variables that had the potential of predicting outcomes on two variables: length of stay and whether the conditions of nursing home residents would deteriorate. Included among the independent or predictor variables was the extent of participation in rehabilitation, diagnosis, current health status, impairment status, and prognosis. Palmore used an analysis technique called *automatic interaction detector,* a statistical procedure that can reveal predictive relationships and the strength of those relationships, to examine the effect of the 22 variables on both outcomes. As you can see, predictive statistics are extremely valuable in that they suggest what might happen in one arena (outcome in nursing home placement), based on knowledge of certain indicators (22 predictive variables).

Let us consider an example of the use of one of these statistics, multiple regression. Multiple regression is used to predict the effect that multiple independent (predictor) variables have on one dependent (outcome or criterion) variable. Multiple regression can be used only with interval-level data. Discriminant function analysis is a similar test used with categorical or nominal data. In the article by Rozier et al.[5] that examined the job satisfaction of occupational therapy faculty, the investigators were interested in predicting the extent to which four variables could predict teaching satisfaction. The four variables were "pleasant environment, much satisfaction with teaching, designing learning experiences and the lack of constraints posed by higher education on the teaching process" [p 163].[5] The multiple regression is an equation based on correlational statistics in which each predictor variable is entered into the equation to determine how strongly it is related to the outcome variable and how much variation in the outcome variable can be predicted by each independent variable. In some cases, a stepwise multiple regression is done, in which the predictors are listed from least related to most related and the cumulative effect of variables is reported. In the Rozier et al.[5] study, it was found that all four predictor variables were important influences on the variance of the outcome variable.

Other techniques such as modeling strategies are frequently used to understand complex system relationships. Suppose you were interested in determining why some persons with chronic disability could live independently and others could not. With so many variables, you might choose a modeling technique that would help you identify mathematical properties of relationships, allowing you to determine which factor or combination of factors could best predict success in independent living. We cannot cover these complex, advanced statistical procedures in a text of this type and refer you to the bibliography for suggestions of where to begin reading about these techniques.

COMPUTER STATISTICAL PROGRAMS

Before the popular use of desk top computers, only huge data sets were analyzed on main frame computers. However, within the past decade, superior hardware and software have made computerized data analysis available to most researchers. There are many statistical analytic programs on the market, some more user friendly than others and some more capable of complex computations than others. Among the most popular and powerful are SPSS (Statistical Package for the Social Sciences, McGraw-Hill, New York) and SYSTAT (Systat, Evanston, Ill.). SPSS is a program that can run many types of statistical computations on large data sets. It is based on BASIC, a computer programming language, and the user must know exact commands to successfully use it. SYSTAT is another powerful program but is menu driven. In SYSTAT, the user may select a statistical computation from a menu and then is prompted with a series of branches to perform the next step. Each has its advantages and disadvantages. There are many other statistical programs, and we advise you to select one that fits your data analytic needs carefully with the assistance of a statistician. Even with the use of computers, it is necessary to have a conceptual understanding of the range of available statistics, both to enhance reading and use of knowledge and to promote the careful and correct use of statistical techniques.

EXERCISES

1. Select a research article that uses statistical procedures.
   a. Determine the level of statistical techniques used.
   b. Ascertain the rationale behind the selection of the specific statistics used.
   c. Did the statistical tests answer the question the investigator initially posed?
   d. Were the tests chosen appropriate for the nature of the data collected, sample size, nature of variables measured?
2. Given the following scores, develop a frequency table and find the mode, median, mean, range, and standard deviation:
   12 35 34 26 26 13 21 22 22 22 24 35 36 37 39 51 23
   42 41 21 21 22 25 26 27 44 42 13 35 43 12

REFERENCES

1. Kachigan SK: *Statistical analysis: an interdisciplinary introduction to univariate and multivariate methods,* New York, 1986, Radius Press.
2. Huck SW, Cormier WH, Bounds WG: *Reading statistics and research,* New York, 1974, Harper & Row.
3. Royeen C, editor: *Clinical research handbook.* Thorofare, NJ, 1989, Slack.
4. Knoke D, Bohrnstedt GW: *Basic social statistics,* Itasca, Ill, 1991, Peacock.
5. Rozier CK, Gilkeson GE, Hamilton BL: Job satisfaction of occupational therapy faculty, *Am J Occup Ther* 45:160–165, 1991.
6. Pilcher D: *Data analysis for the helping professions: a practical guide,* Newbury Park, Calif, 1990, Sage.
7. Mohr L: *Understanding significance testing,* Newbury Park, Calif, 1990, Sage.
8. Henkel RE: *Tests of significance,* Newbury Park, Calif, 1976, Sage.
9. Andrews FM, Klem L, Davidson TN, et al: *A guide for selecting statistical techniques for analyzing social science data,* ed 2, Ann Arbor, Mich, 1981, University of Michigan.
10. Wechsler D: *Wechsler adult intelligence scale—revised manual,* New York, 1981, Psychological Corp.
11. Palmore E: Predictors of outcome in nursing homes, *J Appl Gerontol* 9:172–184, 1990.

18

Analytic Strategies for Qualitative Research

- **Stage 1: Ongoing Analysis**
- **Stage 2: Reporting the Analysis**
- **Examples of Analytic Process**
- **Accuracy and Rigor in Analysis**
- **Summary**
- **Exercises**

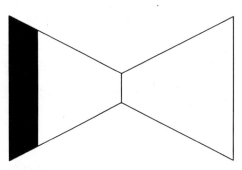

ANALYTIC STRATEGIES
- •Ongoing
- •Formal

The primary purpose of qualitative analysis is to discover and reveal the perspectives or world views of people and the meanings they assign to behaviors and experiences. As we have discussed in previous chapters, the aim is "to comprehend and to illuminate the subject's view and to interpret the world as it appears to him" [p 25].[1] There are many different approaches to achieve this "interior view,"[2] which may include an unstructured analytic strategy, as well as a structured numeric description. These different approaches may be combined and used within any one field study. Also, each naturalistic type of inquiry tends to use a distinct analytic orientation and label the thinking and actions quite differently. We discussed two of these orientations, the emic and etic perspectives, in Chapter 16. Each analytic orientation provides a different understanding and reveals a different aspect of the realities of the field. As Gubrium[3] summarizes:

All areas of study have their methods of procedure. Some are highly structured, being fixed recipes for going about the business of a discipline. . . . Other methods are formed in terms of doing whatever is needed to understand what is being studied, which makes for a looser enterprise. Conceptions of field research have varied accordingly [p 132].

This variation is caused primarily by the varying and distinct philosophical foundations that form the paradigm we label "naturalistic inquiry." It is also a consequence of the epistemologic position that there are multiple realities to discover and reveal, not just a singular objective reality that can be measured using prescribed procedures. Thus, the nature of the analytic process in naturalistic inquiry is necessarily ever changing and evolving as a consequence of the phenomena of the particular field of study.

We cannot possibly cover and do justice to the range of analytic strategies and approaches to fieldwork in this text, and we refer you to a number of excellent books listed in the bibliography. The following discussion is designed to give you an overview of some basic principles that underlie the general thinking and action processes of researchers involved in the naturalistic paradigm.

A cautionary note: the analytic process in naturalistic inquiry is dynamic. It is an intimate part of the information-gathering process that we discussed in Chapter 15. As such, it is contextual and arises from the situation. Analysis of qualitative data is not based on the linear, step-by-step procedures that characterize analysis along the experimental-type continuum or that are implicit in the presentation in this chapter. However, we believe that as a student of research, it is helpful to conceive of the qualitative analytic process as occurring in two overlapping stages, each involving discernible thinking and action processes. We discuss these processes in a linear fashion as a heuristic or instructional device only. We believe such a discussion is informative and can alert you to the concerns and challenges that underlie the dynamic, changing, and evolving analytic structure of qualitative research. However, once you are in the field, the experiences of trying to make sense of what you observe and hear, or "learning the ropes,"[2] do not necessarily lend themselves to a cookbook recipe, step-by-step format.

Table 18–1 shows the analytic process for qualitative inquiry. The first stage we discuss involves an ongoing analytic data–gathering process that occurs while the investigator is in the field. The second analytic stage occurs when the

TABLE 18-1.

Query

Stage 1: Data Collection ↔ Ongoing Analysis
- Inductive reasoning
- Category development
- Taxonomic analysis
- Themes and meanings

↓

Stage 2: Formal Analysis
- Self-reflection
- Interpretation
- Report writing

---

researcher leaves the field and formally writes up the analysis in reportable form.

STAGE 1: ONGOING ANALYSIS

Qualitative research is often referred to as an iterative process. That is, data analysis occurs immediately as one enters the field and is ongoing. Throughout the investigator's engagement in the field, the analysis incrementally builds on ideas as they are generated. Analysis forms the basis from which subsequent field decisions are made regarding whom to interview next, what to observe, or which piece of information to explore further. Ongoing activity involves checking out initial impressions and field-generated hunches through continued review and analysis of each datum that is collected and layer of meaning that is revealed. Thus, analysis of field notes, observations, and transcriptions is an ongoing process that informs and, in turn, is informed by the continual interaction in the field and by collection of data.

Initially, after each interview or observational period, it is important for the investigator to transcribe notes or audiotapes.[4] These notes are then reflected on and used to formulate and reformulate questions and hypotheses, to explore initial investigator perceptions, biases, or opinions, and to group information into meaningful categories that describe the phenomenon of interest. This type of active analysis is especially critical in the early stages of fieldwork to reframe the initial query and set limits as to who and what is investigated.

One can consider this ongoing analytic process as involving four very basic thinking and action processes: inductive thinking, developing categories, developing taxonomies, and discovering meaning and underlying themes. In naturalistic research, these processes are not neat, separate entities that occur in sequence at a particular time in fieldwork. Rather, they are ongoing, overlapping processes that become refined throughout the data-gathering effort. However, in this chapter, we separate and discuss each aspect to understand the underlying framework that guides the qualitative researcher.

Inductive Thinking Process

Thinking about each aspect of the data gathered is a crucial component of the analytic work of qualitative investigators. In describing data analysis in ethnographic research, David Fetterman[5] points out that:

> *The best guide through the thickets of analysis is at once the most obvious and most complex of strategies: clear thinking. First and foremost, analysis is a test of the ethnographer's ability to think—to process information in a meaningful and useful manner [p 88].*

As we have discussed in previous chapters, naturalistic research is based on an inductive thinking process. It is this thinking process that is key to the entire analytic approach to qualitative information. Glaser and Strauss[6] describe the inductive thinking process as one that begins inductively with a hunch or idea. The investigator then explores information or behavioral actions and formulates a working hypothesis. The working hypothesis is then examined in the context of the field to see whether it fits. The investigator then works somewhat deductively and tries to draw implications from the hypothesis as a way of verifying its accuracy. This process characterizes the thinking and actions of the investigator throughout the field experience.

Fetterman[5] refers to this process somewhat differently and labels it "contextualization," or the placement of data into a larger perspective. While in the field, the investigator continually strives to place each piece of data into a context to understand the bigger picture or how parts fit together to make the whole.

Developing Categories

In Chapter 16 we discussed the volume of information that is gathered in the course of fieldwork. The need to manage the immense amount of data collected in the field occurs very quickly. Just think about when you have had to review a large body of literature for a course in college. You first went to the library to obtain information from the literature. As we have all experienced, the notes that you took on your readings soon became overwhelming unless you began to organize them. Without finding a way to organize and make sense of the information you obtained, the task of communicating knowledge in a meaningful way was virtually impossible. The same principle applies for the qualitative researcher. The researcher must find a way to analytically organize and make sense of the information as it is being collected. One of the first meaningful ways in which the investigator begins to organize information is to develop categories. This action process represents a major step in naturalistic analysis. As Wax[7] describes:

> *The student begins "outside" the interaction, confronting behaviors he finds bewildering and inexplicable: the actors are oriented to a world of*

meanings that the observer does not grasp . . . and then gradually he comes to be able to categorize peoples (or relationships) and events [p 325].

As we have discussed before, in naturalistic inquiry the researcher enters the field to see and understand without imposing concepts, labels, categories, or meaning in an a priori fashion. Categories emerge from researcher-field interactions and initial information that is obtained and synthesized. These preliminary categories become tools to sort and classify subsequent information as it is received.

Spradley and McCurdey[8] describe this process as the classification of objects according to their similarities to and differences from other objects. Categories such as "student," "provider," or "client" make it easier to anticipate the behavior of individuals within these groupings and to begin to identify the cultural rules that govern their behavior. As they point out, categories are basic cultural elements and enable people to organize experience. Categories are social inventions in that there is no natural way in which objects are grouped. These groupings change from culture to culture. Take, for example, the way in which colors are classified by different cultures. Where we have only one basic category for the color "white," Alaskans refer to numerous categories. This difference is not surprising, considering the prevalence of "white" in the natural environment in Alaska. Categories are therefore embedded within a cultural context and reveal, at the most basic level, the nature of the way in which a culture classifies objects and its system of meaning.

How do you find or identify categories? The ethnographer, for example, looks for commonalities among an array of different objects, experiences, or events. Agar[5] suggests that recurrent topics are prime candidates for categories to code. Consider the classic ethnography of street corner life that we have discussed throughout the book. Liebow,[9] in *Tally's Corner*, examined similarities and differences in the way in which his informants performed their life roles.

As the data collection activity proceeds, the naturalistic investigator uses the original categories as a basis for analyzing new data. Codes are assigned based on these categories, and new data are either classified into existing categories or serve to modify or create new categories to accurately depict the phenomenon of interest. Lofland and Lofland[10] suggest that the researcher "files" data by placing them into categories based on characteristics that the data may commonly share. The researcher decides on the filing scheme and considers both descriptive and analytic cataloging. Data placed in the descriptive categories answer "who," "what," "where," and "when" queries without interpretation. Analytic categories answer "how" and "why" queries.

The coding of categories involves a repeated review and examination of the narrative. The investigator then assigns codes using any number of methods, depending on one's own personal style of working. For example, some researchers literally cut and paste their data into the categories in which they fit, whereas others use filing boxes, index cards, and dividers. New computer programs for qualitative research such as Zyindex and Ethnograph often facilitate this coding process for very large data sets. These programs automatically assign codes to data based on key words the investigator identifies within the narrative. Some researchers prefer to use word processors to organize data. Depending on one's coding scheme and scope of the project, these programs may make it easier to cut and paste or to copy data into multiple categories.

For example, the researcher might program the computer to automatically assign the code of "self-care" to any datum that has the term "bathing" in it. These computer programs are similar in nature to the organizing and cataloging systems used in libraries. As we discussed in Chapter 5, when searching the literature, you select key words and enter them into the computer. Because librarians have already developed a sophisticated index system, the computer is able to find categories of literature with no more information than your key words. In using computer programs for categorizing naturalistic data, you take the place of the librarian by creating your own indexing and cataloging system. You select the key words based on themes that arise from your data set and then program the computer to automatically assign codes based on your key words.

Developing Taxonomies

The development of a taxonomy represents the next level of complexity and organization of information. Taxonomic analysis involves two processes: organization or grouping of similar or related categories into larger categories, and identification of differences between sets of subcategories and larger or overarching categories. In this process, related subcategories are grouped together. So, for example, whales and dogs belong to the larger category of animals, and blocks and dolls belong to the larger category of toys. In a taxonomy, sets of categories are grouped based on similarities. The investigator must uncover the thread or inclusionary criteria that link categories. Taxonomic analysis is therefore an analytic procedure that results in an organization of categories and describes their relationships. In a taxonomic analysis, the focus is on identifying the relationship between wholes and parts.

Discovering Meaning and Underlying Themes

Agar[4] suggests that in an ethnography it is from the "simple" process of first establishing topics, categories, and codes that: "you begin building a map of the territory that will help you give accounts, and subsequently begin to discuss what "those people" are like" [p 105].

So far, we have spoken about developing categories and taxonomies in a somewhat unidimensional way. But one of the main purposes of naturalistic research is to understand how each observation or part fits into the whole to make sense out of the layers of meaning and the multiple perspectives that compose field experience. The major inductive method to accomplish the task is to search for relationships among categories and to reveal the underlying meaning in categories and their components beyond what is immediately visible.

According to Glaser and Strauss,[6] the researcher engages in the thinking process of "integrating" in which he or she finds relationships among categories and further looks for overlap, exclusivity, or hidden meaning among categories. In their grounded theory method, Glaser and Strauss use the term "theoretical sensitivity" to refer to the researcher's sensitivity and ability to detect and give meaning to data, to go beyond the obvious and to recognize what is important in the field.[11, 12]

In Spradley and McCurdey's[8] terms, the ethnographer looks for patterns of thought and behavior by examining repeated actions. Categories and taxonomies are compared, contrasted, and sorted until a discernible thought or behavioral pattern becomes identifiable and the meaning of that pattern is revealed. As exceptions to rules emerge, variations on themes are detectable. As themes are developed based on abstractions (categories and taxonomies), they are examined in depth in light of ongoing observations.

These rigorous methods of inductive analysis allow the researcher to uncover the multiple meanings and perspectives of individuals and to develop complex understandings of their experience and interactions.

Savishinsky,[13] in his study of the culture of a nursing home, notes the multiple levels of insights that occur in naturalistic inquiry and the way in which new data can influence previously held notions:

> *The lives of the residents were absorbing because beneath the deceptively simple style in which they told their stories, there was often depth of passion, or the moral twists of fable, or one small detail which transformed the meaning of all the other details [p 69].*

STAGE 2: REPORTING THE ANALYSIS

We have seen how analysis is an active component of the data collection process and helps shape each aspect of field work. Analysis also continues, however, once the investigator formally leaves the field and has completed data collection. In this second more formal analytic stage, the investigator enters an intensive analytic and writing effort. The main objective is to consolidate one's understanding and impressions by writing a final manuscript or report. This writing effort involves, in part, self-reflection and forces the researcher to pull it all together, to move beyond the data and interpret. The ease with which this step is done is often dependent on how well the investigator initially organizes and cross-references the voluminous records and field notes that are obtained.

This more formal analytic stage involves reexamining materials and reflecting on and refining categories, taxonomies, meanings, and interpretations. The investigator must also purposely and carefully select quotes and examples to illustrate and highlight each aspect of the interpretation. Selections are carefully made to assure adequate representation of the interpretation or meaning the investigator wishes to convey and to remain true to the voice that is being quoted and the context in which the narrative occurred.

In the final interpretation, the investigator moves beyond the data to suggest a deeper understanding of the field through theory development or an explication of the meanings and general principles that emerge regarding the phenomena of interest.

EXAMPLES OF ANALYTIC PROCESS

Researchers use the basic analytic processes we have discussed somewhat differently and often label these activities distinctly. One of the most formal and systematic analytic approaches has been developed by Glaser and Strauss[6] for the purpose of developing theory. Here we only highlight aspects of grounded theory to demonstrate one form of a rigorous qualitative analytic method.

Grounded Theory

Glaser and Strauss[6] have suggested very specific procedures that are used widely in qualitative analysis. Grounded theory systematizes the inductive incremental analytic process and the continuous interplay between previously collected and analyzed data and new information. The authors label their method the *constant comparison method.* In the constant comparative method, as information is obtained, it is compared and contrasted to previous information to inductively fit all the pieces together into a bigger puzzle. Patterns emerge from the data set, which are then coded. Data filing occurs by categorizing and coding. In this method, the researchers not only look for themes to emerge but also code each piece of raw data according to the categories in which they belong.

Codes are initially "open," which refers to a "process of breaking down, examining, comparing, conceptualizing and categorizing data" [p 61].[12] "Axial coding" then occurs, which refers to "a set of procedures whereby data are put back together in new ways after open coding by making connections between categories" [p 96].[12] "Selective coding" occurs next, which is the "process of selecting the core category, systematically relating it to other categories" [p 116].[12] Codes reflect the similarities and differences among themes and continue to test the category system through analysis of each datum and categorical assignment as it is collected. In this way the analysis is grounded in and emerges from each and every piece of datum. Theory emerges from the data, is intimately linked to the field reality, and reflects a synthesis of the information gathered.

This method is perhaps one of the most systematic and procedure-oriented processes along the continuum of naturalistic inquiry. In various books, listed in the bibliography, Strauss and Glaser, together, individually, and with other authors, walk the researcher systematically through each analytic step of coding and categorizing and use a prescribed language to identify each procedure and task.

Using grounded theory, one group of researchers[14] describe their analytic approach to category formation in a study of family empowerment in health care situations:

> *The data were coded and analyzed using the constant comparison methods of grounded theory developed by Glaser and Strauss (1967). The emerging theory was expanded and densified by reduction, selective sampling of the literature and selective sampling of the data. A group of peers in a grounded theory seminar reviewed the developed categorical processes for fit with the data. Emerging concepts were discussed with participant families and further modified and refined until the investigators, the informants, and the peer group found them valid [p 83].*

Field Study Designs

Most field studies use a range of analytic approaches, depending on the purpose and nature of the study. They may borrow techniques from grounded theory or use a more general thematic analysis, depending on the particular philosophical stance and analytic orientation. To illustrate one way in which a field study analysis may be conducted from beginning to end, we present DePoy and Archer's[15] research, which was designed to discover the meaning of quality of life and independence to nursing home residents.

First, because of the limited theory that explained quality of life and the meaning of independence from the perspective of elderly residents themselves, DePoy and Archer[15] initially chose a qualitative approach to address their research query. As the primary data collector, Archer began by examining the "social situation" of a nursing home in rural New England, or what Spradley and McCurdy[8] refer to as the place where the research takes place. Archer recorded her data by initially separating her observational log into two categories: the physical environment and the human environment. This organizational framework is based on an environmental competence model that envisions a person's competence as influencing and, in turn, being influenced by aspects of the physical and social surroundings in which behaviors occur.

Archer answered three basic queries with this organizing scheme: What bounds the environment in which the residents live? What physical characteristics does the environment display? Who and what types of interactions occur within the physical environment? Data were collected and preliminary analyses involved determining similarities between objects to derive descriptive categories. Once Archer was able to collect descriptive data about the environment and who was in it, she and DePoy sampled segments of the data set to examine relationships and types of interactional patterns that occurred within the environment. After reading and rereading observational notes, they identified a pattern of the daily schedule that was followed by both residents and formal providers in the institution. This schedule was conceived as a category of its own, and Archer returned to the field to observe specific times of the day to elaborate or broaden the meaning of this category. For example, she observed morning care, meal times, and visiting hours. This set of observations provided her with greater insight as to why people acted in the way that they did during those times and the meaning of the activity for those individuals. Thus, the initial query became reformulated to "Why do they do what they do, and what is the meaning of the activity?"

In addition to initially developing descriptive categories, DePoy and Archer[15] used constant comparison[6] and taxonomic analysis.[8, 16] When the constant comparison approach was used, two categories of resident activity initially emerged from the first set of field notes: self-care and rehabilitation. For example, Archer documented in her field notes an observation of a woman dressing herself with the assistance of an occupational therapist. This datum was initially coded as reflecting two categories of self-care and rehabilitation. The coding occurred as a result of constant comparison in which the observation was compared with previous data in each category and was determined to be sufficiently similar to be included in both.

However, as a consequence of comparing new and previous data, discrepancies emerged. This dissonance required reflection and action involving the revision of the initially developed category scheme to depict the nature of the datum. Further along in fieldwork, DePoy and Archer's constant comparisons revealed that the category of self-care was characterized by residents conducting their morning routine with assistance in dressing by the occupational therapist. However, other observations revealed that numerous residents were sitting in their beds during the morning self-care time schedule calling for nursing staff. Did these observations fit into the category of self-care? Because the residents were exercising their right to obtain help from staff, the data could be coded as self-care. However, because residents appeared to need and be dependent on requesting assistance, the data did not seem to fit the concept of independence inherent in the category of self-care. Self-care was thus reconceptualized to reflect a new meaning and was eliminated as a category by itself. The example of the woman dressing with the assistance of the occupational therapist was now coded as "functional relationships" (Fig 18–1), whereas observations of those remaining in their beds was revised as subcategories of "in waiting," and "confinement."

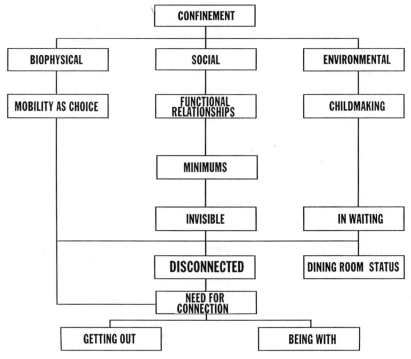

FIG 18–1.
Taxonomy.

As categories emerged from the data set and became clarified by constant comparison, DePoy and Archer examined relationships among them. Figure 18–1 depicts the simple taxonomy of nursing home experience that inductively emerged from the full data set. As you can see, the lines demonstrate bidirectional connections among categories of findings. For example, the category of confinement was broken down into three subcategories: biophysical, social, and environmental confinement. Biophysical confinement was both caused by and resulted in mobility choices that were intimately related to the category of "functional relationships" (relationships between residents and staff based on resident need) and so forth.

ACCURACY AND RIGOR IN ANALYSIS

In naturalistic research, the debate continues as to how to construct standards for conducting and evaluating data gathering and analytic efforts. What are the components of a good or adequate inquiry? How can one determine the accuracy of an investigator's interpretation of field data? Are the experiences of individuals "accurately" represented and analyzed? Or are the themes and taxonomies merely the biases and personal interpretations of an investigator? Do the findings reveal meanings that would be shared by others if conducting the same set of interviews, observations, and analytic orientation?

Lincoln and Guba,[17, 18] in numerous books and articles, have labeled the concern with truth value and accuracy as the *credibility* of the investigator's findings and interpretations. They and other researchers have identified various actions or techniques to enhance the rigor or confidence in the truth value of the findings of field research.

The truth value or credibility and rigor of qualitative research, however, cannot be determined by using the criteria that have traditionally been applied to quantitative research. In reading a qualitative report, the critical reader needs to ask two primary questions. First, to what extent are the bias and personal perspectives of the investigator identified and considered in the data analysis and interpretation? Second, what actions has the investigator taken to enhance the credibility and rigor of the investigation? Here we review six basic actions to enhance accuracy and rigor: triangulation, saturation, member checks, reflection, audit trail, and peer debriefing.

Triangulation

Triangulation, which we also discussed in Chapters 11 and 15, is a basic aspect of data gathering that also shapes data analysis. It is an approach to data analysis in which one source of information is checked against one or more other different types of sources to determine the accuracy of hypothetical understandings. It involves, for example, the comparison of verbal information with written materials or use of observation to confirm or gain additional insight from interview information. By triangulating distinct sources of information, the researcher both verifies an understanding and develops a more comprehensive picture of the phenomenon of interest. As you have already seen in our data collection discussion, it is wise and rigorous to use several strategies to respond to the same query. Doing so not only expands and verifies your data set but also enhances the credibility of the findings. For example, you would be more likely to believe statistics on teenage alcohol consumption if self-report data were verified with observational data and possibly an examination of the number of beer bottles in the trash can where your informants have held their most recent party.

Saturation

Saturation is another indicator that the naturalistic study is rigorous. Saturation refers to the point at which an investigator has obtained sufficient information from which to obtain an understanding of the phenomenon. When the information gathered does not provide any new insights or understandings, it is a signal that fieldwork can be concluded. If you can guess what your respondent is going to do or say in a particular situation, you probably have obtained saturation. As in the case of *Tally's Corner,* Liebow's[9] prolonged engagement in the field assured that the investigator obtained a point of saturation.

However, lengthy emersion in the field may not be practical for service provider researchers who perhaps have limited time and funds to support such an endeavor. DePoy and Archer,[15] for example, were constrained by time and money in collecting data in the nursing home study because of time and resource constraints. They therefore had to use other strategies such as triangulation to heighten the accuracy and rigor of their understandings and interpretations. They also used "random observation" as a strategy to assure saturation. In this technique, the investigators randomly select time throughout the investigative period to increase the likelihood of obtaining a total picture of the phenomenon of interest. DePoy and Archer, for example, selected random times during the week, including morning, afternoon, and night, to observe. In this way they tried to approximate full immersion in the culture of the nursing home. Through random observation, it is theoretically possible to sample the total cycle of the phenomenon of interest.

Member Checks

Member checks are a technique whereby the investigator checks out his or her assumptions with the informants. In the DePoy and Archer study, once the theme of "disconnectedness" emerged from the data set, it was verified with informants by asking them if the investigators' interpretation was accurate. Similarly, Wuerst and Stern[14] verified the meanings that they extracted from their data with the participating families. This type of affirmation decreases the potential for the imposition of the investigator's bias where it does not belong. This technique is very important and is used throughout the data collection process to confirm the truth value or accuracy of one's observations and interpretations as they emerge. The informants' ability to correct the vision of their stories is critical to this technique.

The importance of routinely using member checks is highlighted by Frank's[19] life history of a congenital amputee. In this study, Frank observed several self-care activities of Diane DeVries, a woman with quadrilateral limb deficiencies. After writing up her impressions, Frank asked Ms. DeVries to read the description and comment. This is how Diane DeVries responded to Frank's written observations:

> *Just a few technical corrections. I have a stool which enables me to get from the floor back up to my chair. At that time, my bed just happened to have*

been on the floor. I can get in and out of any bed—from wc [wheelchair] to bed and reverse—from bed to floor (not falling out! HA) but not the reverse.

I balance a glass between my face and arm—if it were my shoulder I would die of thirst. Picture that, and you'll see what I mean. No one needs to place it there. I can pick up a glass, cup, or can and drink it myself.

*I **do** and **can** brush my teeth. I place the brush in my mouth and move it with both arms—for the uppers—with my right arm and tongue for everywhere else. No, I cannot floss.*

*Doors. I can **open** any door as long as I can get to the doorknob, which I turn with my arm and chin. I cannot close a door behind me (except, of course, for a sliding door) [p 641].*

This passage provides great insight into the spirit and personality of Ms. DeVries. It also indicates the gross inaccuracies of the investigator's observational impressions. Some of Frank's initial observations, such as the transfer from chair to bed, were obviously serendipitous. In other words, her observation did not reflect the usual daily routine DeVries typically followed, and thus Frank's conclusions were inaccurate. Frank increased the credibility of her understandings of self-care performance and adaptation to disability by having her informant check and correct the accuracy of her account.

Reflexivity

Although investigator bias cannot be eliminated, it can be identified and examined in terms of its impact on data collection processes and interpretative accounts. The term reflexivity refers to the process of self-examination. In reflexive analysis, the investigator examines his or her own perspective and determines how this perspective has influenced not only what is learned but how it is learned. Through reflection, investigators evaluate how understanding and knowledge are developed within the context of their own thinking processes. Frank, for example, honestly described her biased and initial perspective as she entered into a study of the life of Diane DeVries. She first met DeVries as a student at UCLA in 1976 and explains:

As I observed this woman, . . . I imagined that she lived at home with her parents and would probably never marry or have sexual experiences. In other words, the question arose, "What kind of life could such an individual have in our society?" [p 639].

This initial broad query reflected Frank's limited understanding of the lives of individuals with disabilities. Through the life history approach, Frank's query became reformulated to a quest to understand the process of adaptation to severe congenital disability and how that adaptational process reflected the individual's own sense of cultural normalcy as a member of American society.

A personal diary or method of noting personal feelings, moods, attitudes, and reactions to each step of the data collection process is a critical aspect of the data collection process in naturalistic inquiry. These personal notes form the backdrop from which to understand how a particular meaning or analysis may have emerged at that data collection stage. Fetterman[20] suggests that keeping a personal diary is an effective "quality control" mechanism.

As an insider in the social scene of the nursing home, Archer[15] had to consistently assess how her relationship to the residents and staff influenced her focus and her interpretation of the data.

Krefting's[21] ethnography of 21 persons with moderate head injuries highlights the significance of personal diary keeping as an analytic component to fieldwork. In this study, Krefting used a variety of data collection strategies, including nonstructured interviews, participant observation, and review of artifact materials, to examine the life experiences of persons disabled by head injury and their family members. A major theme that emerged as a life-organizing perspective for persons with brain injury and their family members was the process of "concealment." In the following passage, Krefting[21] discusses how her personal notes reinforced the theme of concealment and solidified this component of her analysis:

> At first I believed that the experience of head injury was "not all that bad," as I noted in my field diary. What I mistook for the real experience was one largely concealed by the disabled persons and their families. It took many months for them to reveal some of the more meaningful but tragic aspects of their lives [p 73].

Audit Trail

Another way to increase rigor is to maintain an audit trail as one proceeds analytically. Lincoln and Guba[17] suggest leaving a path of one's thinking and action processes so that others can clearly follow the logic and manner in which knowledge was developed. In this approach, the investigator is responsible not only for reporting results but also for explaining how the results are obtained. By explaining the thinking and action processes of an inquiry, the investigator allows others to agree or disagree with each analytic decision and to confirm, refute, or modify interpretations.

Peer Debriefing

One technique to ensure that data analysis represents the phenomena under investigation is the use of more than one investigator as a participant in the analytic process. Throughout a co-investigated study, researchers frequently conduct analytic actions independent of one another to determine the extent of their agreement. Synthesis of analysis and examination of areas of disagreement shed additional understandings on the research query. In the case where only one investigator engages in a project, he or she often will ask an external analyzer to function in the capacity of a co-investigator. In some cases a panel of experts or advisors is used to evaluate the analytic process. This form of peer debriefing provides an opportunity for the investigator to reflect on other possible competing interpretations offered in the peer review process and in this way strengthen the legitimacy of the final interpretation. Wuerst and Stern[14] used peer debriefing at multiple time intervals throughout their study to modify and affirm their interpretations.

SUMMARY

We have said that the overall aim of analysis is to search for patterns, be they descriptive or analytic, simple or complex, that emerge from the information obtained. Some researchers seek to generate theory, whereas others aim to reveal, interpret, and communicate the multiple layers of understandings of human experience that emerge in the field. In naturalistic research, data analysis is based on an inductive thinking process that is ongoing and interspersed with the activity of gathering information. For the most part, analysis is an integral component of the data collection process and informs that activity through development and testing of category schema. Through the inductive thinking process, the boundaries of the study become reformulated and defined, and initial descriptive queries are answered. Further data collection efforts are determined by the need to more fully explore the depth and breath of categories and to answer why and how type of questions. As data are obtained to unravel the layers of complexity of a query, they are coded and organized into meaningful categories. The boundaries and meanings of categories are further refined through the process of establishing relationships among categories. Why and how queries are informed through the development of taxonomies that lead to emergent patterns of meaning and interpretations of how observations fit into a larger whole or make the realities of the field.

Gubrium[3] summarizes the analytic process as follows: "Analysis proceeds incrementally with the aim of making visible the native practice of clarification, from one domain of experience to another, structure upon structure" [p 5].

EXERCISES

1. Return to the exercise that you conducted in Chapter 15 where you observed a public place to determine characteristic behavior patterns. Now examine your raw data and search for categories. Based on the categories, code each datum and develop a descriptive taxonomy of behavior.

2. Develop an audit trail for your data collection and analysis activities in the previous inquiry.

3. Give your raw data set from exercise no. 1 to a colleague for analysis and the establishment of an audit trail. When your colleague has completed the task, compare your conclusions. Reconcile any differences by reexamining your data and your audit trails.

REFERENCES

1. Matza D: *Becoming deviant* Englewood Cliffs, NJ, 1969, Prentice Hall.
2. Shaffir WB, Stebbins RA, editors: *Experiencing fieldwork: an inside view of qualitative research,* Newbury Park, Calif, 1991, Sage.
3. Gubrium J: *Analyzing field reality,* Newbury Park, Calif, 1988, Sage.
4. Agar MH: *The professional stranger: an informal introduetion to ethnography,* New York, 1980, Academic Press.
5. Fetterman DL: *Ethnography step by step,* Newbury Park, Calif, 1989, Sage.
6. Glaser B, Strauss A: *The discovery of grounded theory,* Chicago, 1967, Aldine.
7. Wax M: On misunderstanding verstechen: a reply to Abel, *Sociol Soc Res* 51:323–333, 1967.
8. Spradley JP, McCurdy DW: *The cultural experience: ethnography in a complex society,* Prospect Heights, Ill, 1988, Waveland Press.
9. Liebow E: *Tally's corner,* Boston, 1967, Little, Brown.
10. Lofland J, Lofland L: *Analyzing social settings: a guide to qualitative observation and analysis,* ed 2, Belmont, Calif, 1984, Wadsworth.
11. Glaser BG: *Theoretical sensitivity: advanced in the methodology of grounded theory,* Mill Valley, Calif, 1978, Sociology Press.
12. Strauss AL, Corbin JM: *Basics of qualitative research: grounded theory procedures and techniques,* Newbury Park, Calif, 1990, Sage.
13. Savishinsky JS: *The ends of time: life and work in a nursing home,* New York, 1991, Bergen & Garvey.
14. Wuerst J, Stern PN: Empowerment in primary health care: the challenge for nurses, *Qualitative Health Res* 1:80–99, 1991.
15. DePoy E, Archer L: Quality of life in a nursing home, *Topics Geriatr Rehabil* 7:64–74, 1992.
16. Miles MB, Huberman AM: *Qualitative data analysis: a sourcebook of new methods,* Newbury Park, Calif, 1984, Sage.
17. Lincoln YS, Guba EG: *Naturalistic inquiry,* Newbury Park, Calif, 1985, Sage.
18. Guba EG: Criteria for assessing the trustworthiness of naturalistic inquiries, *Educ Commun Technol J* 29:75–92, 1981.
19. Frank G: Life history model of adaptation to disability: the case of a congenital amputee, *Soc Sci Med* 19:639–645, 1984.
20. Fetterman DL: *Ethnography step by step,* Newbury Park, Calif, 1989, Sage.
21. Krefting L: Reintegration into the community after head injury: the results of an ethnographic study, *Occup Ther J Res* 9:67–83, 1989.

19 | Reporting and Disseminating Conclusions

- **Principle for Writing**
- **Writing an Experimental-Type Report**
- **Writing a Naturalistic Report**
- **Writing an Integrated Report**
- **Dissemination**
- **Summary**

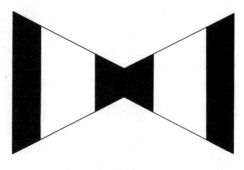

REPORTING

Reporting and disseminating the results of your study are two important action processes in research. Research, in essence, is not research unless it is shared with others who can benefit from it. Any research study may generate one or more reports. Reports are designed to fit the particular avenue or context you choose to disseminate your findings. The method you choose for reporting and disseminating your research is purposeful and driven by the nature of your question or query.

Remember the definition of research we presented in Chapter 1? We specified four criteria to which a study must conform: logic, understandability, confirmability, and usefulness. Reporting and disseminating are two research action processes that also reflect these criteria. However, usefulness is the unique research criterion that cannot be accomplished without telling others about your findings and their meanings.

As we have suggested throughout the book, there are multiple research languages and structures. Likewise, there are multiple ways in which you may choose to present and structure the reporting and disseminating of your research. The reporting language and structure you use must be congruent with the epistemologic framework of your study and fit the purpose for dissemination.

Writing and disseminating a research project is analogous to telling a story. As indicated by Richardson[1]:

> *Whenever we write science, we are telling some kind of story, or some part of a larger narrative. Some of our stories are more complex, more densely described, and offer greater opportunities as emancipatory documents; others are more abstract, more distanced from lived experience and reinscribe existent hegemonies [p 13].*

PRINCIPLES FOR WRITING

Let us now consider the writing action process for both the naturalistic and experimental-type continua. As we indicated, each continuum has a distinct language or set of languages and structures that organize the reporting action process. However, all written research reporting has a common set of principles that can be used to guide this important action process.

Clarity

As stated so clearly by Babbie,[2] writing a report serves little purpose if it cannot be understood. Therefore, an investigator should ensure that a written report is clear and well written.

Purpose

As we have indicated throughout this book, there are multiple purposes that drive the selection of action processes in research. These purposes also structure the nature of the reporting action process.[3] Consider, for example, two differing purposes: publishing and evaluating. Researchers who write for the purpose of publishing their research in scholarly journals must conform to the style and expectations of the journal to be considered for publication. The researcher might also consider writing an article for use by practitioners and thus would present results and interpretations somewhat differently. Or, the researcher who has just completed an evaluation required to assure continued funding for a program might emphasize positive programmatic outcomes and write the research report so that it is consistent with the expectations of the funding agency. There are many purposes for conducting and reporting research, and each purpose should be carefully considered during the report and dissemination action phase.

Audience Specific

Along with purpose, the audience for which the report is written in large part determines how the report will be structured and the degree of specificity that will be included. The audience may vary in areas of expertise and knowledge of research methodology. The important point is to identify your reason for writing a report to a particular audience and to assess the way in which that purpose can be best accomplished for that audience. You need to write in a style that is consistent with the level of understanding and knowledge of the targeted reader.

With these common sense principles in mind, let us now consider the specific writing considerations for each design continuum.

WRITING AN EXPERIMENTAL-TYPE REPORT

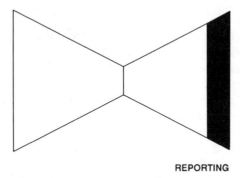

REPORTING

Of no surprise, when an experimental-type research report is written, there is an agreed on language and structure for presentation traditionally used to present experimental-type research. The language used by the experimental-type researcher is "scientific" in nature. That is, the writing is logical and detached from personal opinion. Interpretations are supported by numeric data.

In the experimental-type report there are usually seven sections, ordered as follows:

- Abstract
- Introduction
- Review of literature
- Method
- Results
- Discussion
- Conclusions

Although investigators sometimes deviate from this order of presentation, these seven sections are usually considered the critical components of a scientific report.

Let us examine each section. The *abstract* appears before the full report and briefly summarizes or highlights the major points made in each subsequent section of the report. It includes a statement of purpose, a brief overview of method, and a summary of the findings and implications of the study. The abstract usually does not exceed two paragraphs. Journals usually specify the length of the abstract.

The *introduction* section presents the problem statement, the purpose statement, and a review of the questions that the study will answer. It is in this section that the researcher can address a specific audience and set forth the contribution that the research is intended to make to professional knowledge and practice.

As we discussed in Chapter 5, the *literature review* creates the conceptual foundation for your research. In the introduction of a report, you need to summarize the important literature that addresses the key concepts, constructs, principles, and theory of your study. Refer back to your literature chart or concept matrix to help you organize and focus this section. The degree of detail in the literature review once again depends on the purpose and audience. For example, in a journal article, the researcher usually limits the literature review to the seminal works that precede the study, whereas in a doctoral dissertation all previous work that directly or indirectly informs the research question is included. Some researchers combine the introduction and the literature review into one section. Grinnell[3] has named the combined introduction and literature review "the problem statement."

The *method* section consists of several subsections, including the research question or questions, the population and sample, the instrumentation, the procedures used to conduct the research, and the specified data analytic strategies. The degree of detail and specificity is once again determined by the purpose and audience.

In the *findings,* or *results,* section, the analyzed data are presented. Usually, researchers begin by presenting descriptive statistics of their sample population and then proceed to a presentation of inferential and associational types of statistical analyses. A rationale for each statistical analysis procedure is presented, and the findings are usually explained. However, interpretation of the data is usually not presented in this section. Data may be presented in narrative, chart, graph, or table form. You should be aware that there are accepted formats for presenting statistical analyses, and we suggest that you consult Huck et al. and their other statistical sources that appear in the bibliography for guidance.

The *discussion* section is perhaps the most creative part of the experimental-type research report. In this section, the researcher discusses the implications and meanings of the findings, poses alternative implications, relates the findings to previous work, and suggests the use of the research results. Most researchers include a statement of the limitations of the study in this section as well, although it can also be found in method or findings.

The *conclusion* section is a short summary that draws conclusions about the findings of the study in regard to future directions for research or health care practices.

As you can see, writing an experimental-type report follows a logical, well-accepted sequence that includes seven essential sections. The degree of detail and precision in each section is dependent on the purpose and audience for the report.

WRITING A NATURALISTIC REPORT

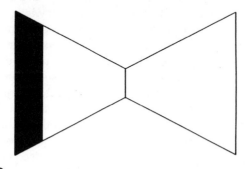

REPORTING

Because there are so many epistemologically and structurally different types of research designs on the naturalistic continuum, there is no single, accepted report writing format. However, there are some basic commonalities among the varied designs that can guide your report writing in this tradition. Unlike experimental-type reporting, naturalistic reporting does not follow a prescribed format with clear expectations for language and structure. Rather naturalistic reporting reflects the complex nature of understanding and uncovering a domain and the multiple levels of experience that are interpreted. Consequently, written reports are based on the narrative derived from data collection, and the researcher often includes his or her personal values and perspectives about the process and outcomes throughout the report.[1]

Although the report may be structured in many different ways, reporting does contain some of the basic elements discussed above for experimental-type research. Using the language of this research tradition, the sections for naturalistic report writing are:

- Critical elements of a report in naturalistic inquiry
- Introduction
- Query
- Research process
- Information
- Analysis
- Meaning and implications

Reporting in naturalistic research may not necessarily follow the sequence outlined, but each report should contain these basic elements.

The *introduction* in naturalistic design contains the purpose for conducting the research. An investigator may include personal purposes, as well as purpose related to the development of professional knowledge.

The *query*, research process, information, and analysis may appear separately or may be combined in one section of the report. As we indicated, these thought and action processes are simultaneous throughout the conduct of naturalistic research and can be presented as such. However, consistent with the principle of clarity discussed earlier, investigators need to select a format that will accomplish a purpose but that will be clear and easily understood by the targeted audience. By now, you probably noticed that literature review was omitted from this list. In naturalistic reporting, we are not forgetting the literature. Rather, consistent with the research tradition, we include research and professional literature along with other types of media in the category of information. Of course, the design will determine the role of research literature in the report. That is, the literature may be presented as a conceptual foundation for the query or as additional evidence for analysis.

Naturalistic research departs from experimental-type reporting in the presentation of information as well. As indicated by Richardson,[1] naturalistic investigators rely heavily on narrative to report their data. It is not uncommon to find quotations from informants woven throughout the narrative text. Visual and numeric display of information, such as taxonomic charts and content analyses, respectively, are also used in naturalistic reporting depending, of course, on the query, the epistemologic foundation, the purpose, and the audience. Let us consider how the reports from two different naturalistic designs might be structured.

1.   Ethnography. — In ethnography, although there are divergent schools of thought, as we indicated earlier in the book, the primary function is to characterize a culture. The report therefore tells a story about a culture through reporting its patterns, roles, and norms. It is not unusual for ethnographies to begin with method. That is, the investigator frequently begins by reporting how he or she gained access to the culture. The report then may be developed chronologically, and literature, other sources of information, and conclusions are interspersed throughout and summarized in the end. Liebow,[4] in his book *Tally's Corner*, exemplifies this structure. Because ethnographies, by virtue of the expansiveness of the domain and the nature of narrative data, are often lengthy, summaries rather than full reports of ethnographic findings may appear in journals.

2.   Phenomenologic Design. — Because the focus in phenomenologic research is on the unique experience of one or of a small group of persons, the report is often written in the form of a story. If specific points about the lived experience of the informants are made, they are fully supported with information from the informants themselves, frequently in the form of direct quotes from the field notes. Because the length of phenomenologic studies varies extensively, they may appear as full-length journal articles, books, or any literary style in between. The reports highlight life experience and its interpretation by those who experience it, minimizing interpretation by the investigator.

As you can see, the nature and purpose of each of these designs differs, as does the reporting format. In both cases, form follows function. In ethnography, the function of the investigator is to make sense of what he or she has observed, whereas in phenomenologic study, the investigator reports how others make sense of their experience. These differences are clearly reflected in the written report.

In summary, naturalistic researchers use a variety of written reporting action processes to share their findings. Although there is a heavy reliance on narrative data, the naturalistic investigator has the option of using numeric and visual representations and other media. All reports contain the same ingredients listed earlier, but the order and the emphasis differ across the naturalistic design continuum.

WRITING AN INTEGRATED REPORT

Because integrated reports have made a relatively recent appearance in the research world, there are not set guidelines for language and format. In addition, the nature and structure of the report depend on the level of integration, the types of designs used, the purpose of the work, and the audience who will receive the report. We do, however, offer two principles that may help you in your writing action process. First, you should include the components of a complete research report (see above). Regardless of the length, organization, or complexity, your report should contain a statement of purpose, review of the literature, methodology section, presentation of findings, and conclusions. Second, because integrated design is not as well established as designs along the experimental-type and naturalistic continua, it is often useful to include an evaluation of the methodology in the report and the justification for the use of each design and the way in which they complement each other.

There is just one other important point that you need to be aware of as you develop written reports and methods for dissemination: the issue of plagiarism. We realize that most researchers who plagiarize do not intend to do so. To avoid this potential and devastating mistake, you need to be aware of the norms for citation and credit. All work done by another person, even if not directly quoted in your work, must be cited. There are many different citation formats used in health and human service research. We refer you to your publication source for the correct format and urge you to become familiar with it. If necessary, have someone else check your work to ensure that you have properly credited other authors.

DISSEMINATION

There are many ways in which you may choose to disseminate your research. Let us look at three popular modes of disseminating.

Sharing Written Reports

Sharing a written report is one of the most common ways of disseminating your findings. As we have already discussed, written reports may take on many formats, depending on the purpose of the report and the audience. Although journal articles are perhaps the most popular outlet for disseminating your written report, there are many other valuable formats. These include research briefs in periodicals and newsletters, summary articles in professional newsletters, local newspapers, university newsletters, and so forth, reports to funding agencies, masters and doctoral theses, chapters in edited books, and full-length books. Once you have written a report, the next action process is publication.

Publishing Your Work

The process of getting your work published is exciting but sometimes frustrating if you are new at it. The first step in thinking about publishing is to ensure that your work is of high quality and meets the rigorous standards for research discussed in previous chapters. However, writing a good report does not ensure publication. We suggest that if you are attempting to publish your first piece of written work, consult with someone who is knowledgeable about the process. It is critical to select a medium that is compatible with the design, purpose, focus, and level of development of your work. Each publication has a set of guidelines and requirements for authors that are usually available in one or more of the yearly issues of the publication. Scholarly and professional journals usually publish instructions for authors at least twice per year. These guidelines include a list of the required format for writing, typing and notation style, and submission processes. For refereed journals, once you submit your work, it is sent out to several (usually three) reviewers for comment and evaluation. This process takes between 3 and 6 months. Do not be surprised if your work is rejected. Many journals have many more submissions than they can publish, and rejection does not mean that your work is poor. It is not unusual for journal editors to suggest alternative journals that might accept your work. If your written report is accepted, more frequently than not it will be accompanied with requested revisions from the reviewers or the editor. After revisions, the work is placed in an issue, and galley proofs are sent to you for your review. These galleys are typeset models of your work. Be vigilant in reviewing them to ensure that no mistakes have occurred in the typesetting process. Then celebrate your work as you read it in publication. You have met the research criterion of "usefulness"!

Sharing Your Research Through Other Methods

Written dissemination is not the only method for sharing your work. There are multiple other outlets, including presentations at professional and scholarly conferences, oral presentations in other forums, continuing and inservice education, collaborative work with colleagues, and many other formats. Sharing your work is not only useful to others but also helps you receive constructive criticism to advance your own thinking and conceptual development.

SUMMARY

Disseminating your research is a major and essential part of the research process. Sharing your work provides knowledge to others to inform practice, tests existing knowledge and theory, develops new knowledge and theory, and ultimately promotes the collective advancement of knowledge in health and human services.

REFERENCES

1. Richardson L: *Writing strategies: reaching diverse audiences,* Newbury Park, Calif, 1990, Sage.
2. Babbie E: *The practice of social research,* ed 6, Belmont, Calif, 1992, Wadsworth.
3. Grinnell RM: *Social work research and evaluation,* Itasca, Ill, 1988, Peacock.
4. Liebow E: *Tally's corner,* Boston, 1967, Little, Brown.

20 | Conclusion

- **Stories From the Field**

Well, here we are at the end of the book, but certainly not, we hope, the end of your involvement in research. The beauty of research is that you learn from doing, and it is a never-ending process. At each step of the process, new, more refined, and complex queries and questions emerge and whet your appetite.

We have introduced you to what some might feel are "heavy" philosophical thoughts, a technical language, and logical ways of thinking and acting. These are your tools for you to now soar and creatively explore that which troubles or challenges you in your practice, daily life, professional experience, or in reading of the literature. As you apply these tools to health and human service–related issues, you will discover both the artistry and science involved in research.

Doing research, like any other human activity, has its low and high points, its tedium and thrill, its frustrations and challenges, its drastic mistakes, and clever applications of research principles. Research is, above all, a thinking process. If you think about what you are doing and reflect on what you did, you can learn from your mistakes and keep refining your skill as an investigator.

We would like to share with you some of our own stories from the research field to highlight the twists and turns of doing research and how human this activity really is.

STORIES FROM THE FIELD

Just Beginning

It was an urban anthropology class, and we were split into groups to conduct urban ethnographies on health practices by different ethnic groups in the city. The big assignment: the Chinese community. The research group? A hippie type, a rock musician, and a topless go-go dancer who dressed the part day and night. The threesome entered the Chinese community, a relatively self-contained 5 by 10 block area of the inner city. As you can imagine, we were quite a sight. How would we ever be able to "enter" the world of this community and engage in passive observation and active participation in community activities? We started out by walking around the area and scoping it out. We made observations of the physical environment and spent some great times eating lunch and dinner in various restaurants. We became Chinese restaurant experts for friends and family.

But no breakthrough. No Chinese families agreed to meet with us or to be interviewed. Young people were curious and asked us lots of questions about ourselves, but their parents remained removed, detached, and unavailable. We caused quite a stir in the community. We split up a couple of times to see if one of us could gain access, but we had no luck. Then one day we happened to pass by a small building with an announcement for a Chinese Political Youth Club. The sign was posted in Chinese and English (a sign of a new acculturated generation?), and we walked directly to the address of the club. Here we were welcomed and engaged in long discussions of the political climate in the university and the dilemmas confronting the community. Bingo! This was a beginning. Although our access to the community remained through the ears, eyes, and thoughts of this radical subgroup, we were able to delineate health practices and health issues as perceived by this group. Certainly this was a lesson in nonreactive research and gaining access.

"I'll Do It for You, Sweetie"

It was my first important research position. I was working with a gerontology research institute and learning the ropes of how to coordinate a large federally funded research grant. I was responsible for putting together a rather lengthy face-to-face interview that included questions generated by myself and the research team and an arsenal of standardized instruments used in gerontology. Finally we were ready for pre-testing. There we were with five older adults, who immediately became bored and uninterested and found the questionnaire tedious and irrelevant. "But, I'll do it for you, sweetie," became the constant song. I was devastated, but was assured by my mentors that this is a common response and that the interview schedule was indeed comprehensive and asked what was intended. This was my first big disillusionment with standardized questionnaires.

In Search of Significance!

Five years of intensive interviewing, data entry, data cleaning, and sophisticated statistical analyses, but where is the significance? Accepting no significance when you want to find statistically significant differences between an experimental and control group can be difficult. Although nonsignificance is as important as a difference, sometimes it is a challenge to get published.[1]

Is Health Care Effective?

In our research class, students are required to develop a research question, or query, and a proposal that describes how they intend to answer the problem. Our most frustrating but popular "research question" is the one posed by many beginning students in research, "Is nursing effective?," or "Is occupational therapy effective?"

Is this a researchable question? Can you tell us what is wrong with the way it is posed?

Elevator Insight

The values of occupational therapy students in a particular program were studied by administering a values test to incoming students, first-year students, and second-year students. It was hypothesized that a change in values would occur, which would indicate that students had acquired the professional values that were an intricate component of the curriculum. After the test was administered, some of the students were talking about it on the elevator without realizing that the investigator was present. The students discussed the content of the questions, were comparing responses, and hoped that they had answered them correctly! The investigator realized at that incredible, serendipitous moment, that the students had answered the questions as they thought they should. They thought it was a test with only one set of correct answers. Their responses did not reflect how they actually felt or believed but what was socially and professionally desirable.[2]

Wow, You Got It!

Your heart starts to beat, you feel a rush; it all clicks, falls into place; you uncovered a pattern, finding something really striking—and it's significant. It is right before your eyes. It has the potential of making an impact on how professionals practice, how clients feel[3-5]! Wow, you got it! You feel great.

It is so important; you have to do it again.

REFERENCES

1. Gitlin LN, Lawton MP, Landsberg LW, et al: In search of psychological benefits: exercise in healthy older adults, *J Aging Health* 4:174–192, 1992.
2. DePoy E, Merrill SC: Value acquisition in an occupational therapy curriculum, *Occup Ther J Res* 8:259–274, 1988.
3. Posner J, Gorman KM, Gitlin LN: Effects of exercise training in the elderly on the occurrence and time to onset of cardiovascular disease, *J Am Geriatr Soc* 38:205–210, 1990.

4. Burke JP, DePoy E: Emerging view of mastery, excellence and leadership in occupational therapy practice, *Am J Occup Ther* 45:1027–1032, 1991.
5. Gitlin LN, Corcoran M: Expanding caregiver ability to use-environmental solutions for problems of bathing and incontinence in the elderly with dementia, *Technol Disabil* 2:12–21, 1993.

Glossary

abstract: Part of the research report that appears before the full report and briefly summarizes or highlights the major points made in each subsequent section.

abstraction: Symbolic representation of shared experience.

accidental sampling: See convenience sampling.

analysis of covariance: A statistical technique that removes the effect of the influence of another variable on the dependent or outcome variable.

analysis of variance: Parametric statistic used to ascertain the extent to which significant group differences can be inferred to the population.

artifact review: Data collection technique in which the meaning of objects in their natural contexts is examined.

Associational statistics: A set of procedures designed to identify relationships between multiple variables and to determine whether knowledge of one set of data allows inference or prediction of the characteristics of another set of data.

audit trail: A path of one's thinking and action processes so that others can clearly follow the logic and manner in which knowledge was developed.

axial coding: A set of procedures whereby data are put back together in new ways after open coding by making connections between categories (see citation in Chapter 18).

bimodal distribution: Distribution in which two values occur with the same frequency.

case study: A detailed, in-depth description of a single unit, subject, or event.

chi-square: Nonparametric statistic used with nominal data to test group differences.

cleaning data: An action process whereby the investigator checks the inputted data set to ensure that all data have been accurately represented.

cluster sampling: A random sampling technique in which the investigator begins with large units, or clusters, in which smaller sampling units are contained and then randomly selects elements from those clusters.

code book: A key to record the line and column placement of each observation.

coding of categories: A repeated review and examination of the narrative in naturalistic data analysis.

cohort study: Research examining specific subpopulations as they change over time.

concept: Symbolic representations of an observable or experienced referent.

conceptual definitions: Definitions that stipulate the meaning of concepts or constructs with other concepts or constructs.

concept matrix: A two-dimensional organizational system that presents all information that you have reviewed and evaluated.

concurrent validity: The extent to which an instrument can discriminate the absence or presence of a known standard.

confounding variable: See intervening variable.

constant comparison: Naturalistic data analysis technique in which each datum is compared and contrasted to previous information to inductively fit all the pieces together into a bigger puzzle.

construct: Symbolic representation that does not have an observable or directly experienced referent in shared experience.

construct validity: The fit between the constructs that are the focus of the study and how those constructs are operationalized.

content validity: The degree to which an indicator seems to agree with a validated instrument measuring the same construct.

contextualization: The placement of data into a larger perspective (see citation in Chapter 18).

contingency table: A two-dimensional frequency distribution primarily used with categorical data.

continuous variables: Variables that take on an infinite number of values.

control: The set of action processes that direct or manipulate factors to achieve an outcome.

control group: The group in experimental-type research that does not receive the experimental condition.

convenience sampling (or accidental, opportunistic, or volunteer, sampling): Involves the enrollment of available subjects as they enter the study until the desired sample size is reached.

correlational analysis: Method of determining relationships among variables.

counterbalance design: Experimental-type design that determines the combined effects of two or more interventions and the effect of order on study outcomes through using crossover.

credibility: Truth value and accuracy of findings in naturalistic inquiry.

criterion validity: Correlation or relationship between a measurement of interest and another instrument or standard that has been shown to be accurate.

critical theory: Philosophical/political/sociologic school of thought that seeks to understand human experience as a means to change the world.

critical value: Numerical value indicating how high the sample statistic must be at a given level of significance to reject the null hypothesis.

Cronback's alpha: Statistical procedure used to examine the extent to which all the items in the instrument measure the same construct.

crossover: The reversal of experimental or comparison conditions, used in counterbalance design.

cross-sectional studies: Studies that examine a phenomenon at one point in time.

data reduction: Procedures used to summarize raw data into a more compact and interpretable form.

data set: The set of raw numbers generated by experimental-type data collection.

deductive reasoning: Reasoning based on a cognitive process whereby a general principle or belief is accepted as true and then applied to explain a specific case or phenomenon. This approach in research involves "drawing out," or verifying, what is already accepted as true.

degrees of freedom: The number of values that are free to vary.

dependent variable: The presumed effect of the independent variable.

descriptive research: Research that yields descriptive knowledge of population parameters and relationships among those parameters.

descriptive statistics: Procedures used to reduce large sets of observations into more compact and interpretable forms.

discrete variables: Variables with a finite number of distinct values.

effect size: The strength of differences in the sample values that the investigator expects to find.

embedded case study: Case study design in which the cases are conglomerates of multiple subparts or are subparts themselves placed within larger contexts.

emic perspective: The "insider's," or informant's way of understanding and interpreting experience.

endogenous research (or participatory action research): Naturalistic design in which the research is "conceptualized, designed and conducted by researchers who are insiders of the culture, using their own epistemology and their own structure of relevance" (see citation in Chapter 10).

ethnography: A naturalistic research design concerned with the description and interpretation of cultural patterns of groups and understanding the cultural meanings people use to organize and interpret their experiences.

etic perspective: Outsider's view that reflects the structural aspects of the field.

experimental group: The group in experimental design that receives the experimental condition.

explanatory research: Research designed to predict outcomes.

exploratory research: Studies conducted in natural settings with the explicit purpose of discovering phenomena, variables, theory, or combination thereof.

ex post facto design: Nonexperimental design in which the phenomena of interest have already occurred and cannot be manipulated in any way.

external validity: The capacity to generalize findings and develop inferences from the sample to the study population.

extraneous variable: See intervening variable.

factorial design: Experimental-type design that investigates the effects of either two or more independent variables (X_1 and X_2) or the effects of an intervention on different factors or levels of a sample or study variables (or both).

field study: Research conducted in natural settings.

focus group design: A naturalistic design that uses small group process to facilitate data collection and analysis.

frequency distribution: The distribution of values for a given variable and the number of times each value occurs.

full integration: Method that integrates multiple purposes and thus combines design strategies from different paradigms so that each contributes knowledge to the study of a single problem to derive a more complete understanding of the phenomenon under study.

grounded theory: A method in naturalistic research that is primarily used to generate theory using primarily the inductive process of constant comparison.

Guttman scale: A unidimensional or cumulative scale in which the researcher develops a small number of items (four or seven) that relate to one concept and arranges them so that endorsement of one item means an endorsement of those items below it.

heuristic research: "A research approach which encourages an individual to discover, and methods which enable him to investigate further by himself" (see citation in Chapter 10).

history: The effect of external events on study outcomes.

holistic case study: Case study design in which the unit of analysis is seen as only one global phenomenon.

hypothesis: A statement about the expected relationships between two or more concepts that can be tested. Hypotheses may be direction or nondirectional.

independent variable: The presumed cause of the dependent variable.

inductive reasoning: Human reasoning that involves a process in which general rules evolve or develop from individual cases, or observation of phenomenon.

inference: The extent to which the samples reflect the population both at pretest and post-test times.

inferential statistics: Statistics used to draw conclusions about population parameters based on findings from a sample.

integrated design: Design used to strengthen a study by selecting and combining designs and methods from both paradigms so that one complements the other to benefit or contribute to an understanding of the whole.

interaction effect: Changes that occur in the dependent variable as a consequence of the combined influence or interaction of taking the pretest and participating in the experimental condition.

interactive effects: The combined effects that several variables have on another variable.

internal validity: The ability of the research design to accurately answer the research question.

inter-rater reliability: Reliability test involving the comparison of two observers measuring the same event.

interrupted time series: Repeated measurement of the dependent variable both before and after the introduction of the independent variable.

interval numbers: Numbers that share the characteristics of ordinal and nominal measures but also have the characteristic of equal spacing between categories.

intervening variables (or confounding or extraneous variables):   Those phenomena that have an effect on the study variables but are not necessarily the object of the study.

judgmental sampling: See purposive sampling.

Kuder-Richardon formula (K-R 20):   Statistical procedure used to examine the extent to which all the items in the instrument measure the same construct.

laboratory study: Research implemented in controlled environments.

level 1 questions: Experimental-type questions that describe phenomena.

level 2 questions: Experimental-type questions that explore relationships among phenomena that have already been studied at the descriptive level.

level 3 questions: Experimental-type questions that ask about a cause and effect relationship between two variables to test knowledge or the theory behind the knowledge.

level of significance: The probability that defines how rare or unlikely the sample data must be before the researcher can reject the null hypothesis.

life history research: Naturalistic design that reveals the nature of an individual life over time.

Likert scale: A type of scale most frequently scored on a 5- or 7-point range indicating a subject's level of positive or negative response to an item.

logical positivism: A philosophical school of thought characterized by belief in a singular, knowable reality that exists separate from individual ideas, reductionism, and based in deductive reasoning.

longitudinal study: Research design in which data collection extends over long periods of time.

main effects: The direct effect of one variable on another.

manipulation: The action process of maneuvering the independent variable so that the effect of its presence, absence, or degree can be observed on the dependent variable.

mean: Average score.

median: Point in a distribution at which 50% of the cases fall above and below.

measurement: The translation of observations into numbers.

measure of variability: The degree of dispersion among a set of scores.

measures of central tendency: Numerical information regarding the most typical or representative scores in a group.

member check: A technique whereby the investigator "checks out" his or her assumptions with the informants.

mixed method design: The combination of multiple strategies and or designs to answer a series of research questions.

mode: Value that occurs most frequently in a data set.

mortality: Subject attrition or dropping out of a study before its completion.

multiple case study: Case study design in which more than one study on single units of analysis are conducted.

multiple comparisons: Statistical tests used to determine which group is greater than the others.

multiple regression: An equation based on correlational statistics in which each predictor variable is entered into the equation to determine how strongly it is related to the outcome variable and how much variation in the outcome variable can be predicted by each independent variable.

networking: See snowball sampling.

nominal numbers: Numbers used to name attributes of a variable.

nonequivalent comparison group design: Quasi-experimental design in which there are at least two comparison groups but subjects are not randomly assigned to these groups.

nonexperimental designs: Experimental-type designs in which the three criteria for true experimentation do not exist.

nonparametric statistical procedures: Statistical formulas used to test hypotheses when (1) normality of variance in the population is not assumed, (2) homogeneity of variance is not assumed, (3) the data generated from measures are ordinal or nominal, and (4) sample sizes may be small.

nonprobability sampling: Sampling in which nonrandom methods are used to obtain a sample.

null hypothesis: Hypothesis of no difference.

one-shot case study: Pre-experimental design in which the independent variable is introduced and the dependent variable is then measured in only one group.

operational definition: An operational definition of the concept reduces the abstract to a concrete observable form by specifying the exact procedures for measuring or observing the phenomenon.

opportunistic sampling: See convenience sampling.

ordinal numbers: Numeric values that assigns an order to a set of observations.

panel study: Similar to cohort design except the *same* set of people are studied over time.

parametric statistics: Mathematical formulas that test hypotheses based on three assumptions: (1) the samples come from populations that are normally distributed, (2) there is homogeneity of variance, and (3) the data generated from the measures are interval level.

participant observation: A naturalistic data collection strategy in which the researcher takes part in the context under scrutiny.

participatory action research: See endogenous research.

passive observation designs: Nonexperimental designs used to investigate phenomena as they naturally occur and to ascertain the relationship between two or more variables.

Pearson product-moment correlation: Statistic of association using interval level data and yielding a score between -1 and $+1$.

peer debriefing: The use of more than one investigator as a participant in the analytic process, followed by reflection on other possible competing interpretations of the data.

phenomenology: Naturalistic inquiry that aims to uncover the meaning of how humans experience phenomena through the description of those experiences as they are lived by individuals.

population: A group of persons, elements, or both that share a set of common characteristics that are defined by the investigator.

post hoc comparisons: Statistical tests used to determine which group is greater than the others.

post-test: Observation or measurement subsequent to the completion of the experimental condition.

practice research: The investigation of human experience within the context of health care and human service institutions agencies or settings.

predictive validity: The extent to which an instrument can predict or estimate the occurrence of a behavior or event.

pre-experimental designs–experimental-type designs: Designs in which two of the three criteria necessary for true experimentation are absent.

pre-test: Observation or measurement before the introduction of experimental condition.

principles: See propositions.

probability sampling: Sampling plans that are based on probability theory.

problem statement: A statement that identifies the phenomenon to be explored and why it needs to be examined or why it is a problem or issue.

propositions (or principles): Statements that govern a set of relationships and give them a structure.

prospective studies: Studies that describe phenomena, search for cause and effect relationships, or examine change in the present or as the event unfolds over time.

purposive sampling (or judgmental sampling): The deliberate selection of individuals by the researcher based on certain predefined criteria.

quasi-experimental design: "Experiments that have treatments, outcome measures, and experimental units, but do not use random assignment to create comparison from which treatment caused change is inferred. Instead, the comparisons depend on non-equivalent groups that differ from each other in many ways other than the presence of the treatment whose effects are being tested" [see citation in Chapter 8].

query: A broad statement that identifies the phenomenon or natural field of interest in naturalistic research.

quota sampling: A nonrandom sampling technique in which the investigator purposively obtains a sample by selecting sample elements in the same proportion that they are represented in the population.

random error: Errors that occur by chance.

randomization: The selection or assignment of subjects based on chance.

range: The difference between the highest and lowest observed value in a collection of data.

ratio numbers: Numbers that have all the characteristics of interval numbers but have an absolute zero point as well.

reflexivity: The process of self-examination.

regression: A statistical phenomenon in which extreme scores tend to regress or cluster around the mean (average) on repeated testing occasion.

reliability: The stability of a research design.

research: Multiple, systematic strategies to generate knowledge about human behavior, human experience, and human environments in which the thought and action process of the researcher are clearly specified so that they are (1) logical, (2) understandable, (3) confirmable and (4) useful.

research design: The plan or blueprint that specifies and structures the action processes of collecting, analyzing, and reporting data to answer a research question.

research question: A clear question that guides an experimental-type investigation. The question is concise and narrow and establishes the boundaries or limits as to what concepts, individuals, or phenomena will be examined.

retrospective studies: Studies that describe and examine phenomena after the fact or after the phenomena have occurred.

sample: The subset of the population that participate or are included in the study.

sampling error: The difference between the values obtained from the sample and that actually exist in those in the population.

sampling frame: A listing of every element in the target population.

saturation: The point at which an investigator has obtained sufficient information from which to obtain an understanding of the phenomena.

scales: Tools for the quantitative measurement of the degree to which individuals possess a specific attribute or trait.

semantic differential: Scaling technique in which the researcher develops a series of opposites or mutually exclusive constructs that ask the respondent to give a judgment about something along an ordered dimension, usually of 7 points.

significance: The extent to which group differences are a function of chance.

simple random sampling (SRS): A probability sampling technique in which a sample is randomly selected from a population.

single case study: Designs in which only one study on a single unit of analysis is conducted.

snowball sampling (or networking): Obtaining a sample through asking subjects themselves to provide the names of others who may meet study criteria.

Solomon four-group design: True experimental design that combines the true experiment and the post-test-only design into one design structure.

Spearman rho: Statistic of association using ordinal level data and yielding a score between -1 and $+1$.

split-half technique: Reliability technique in which instrument items are split in half and a correlational procedure is performed between the two halves.

standard deviation: An indicator of the average deviation of scores around the mean.

static group comparison: Pre-experimental design in which a comparison group is added to the one-shot case study design.

statistical conclusion validity: The power of one's study to draw statistical conclusions.

statistical power: The probability of identifying a relationship that exists or the probability of rejecting the null hypothesis when it is false or should be rejected.

stratified random sampling: Sampling technique in which the population is divided into smaller subgroups, or strata, from which sample elements are then chosen.

survey designs: Nonexperimental designs used primarily to measure characteristics of a population, typically conducted with large samples through mail questionnaires, telephone, or face-to-face interview.

systematic random sampling: Sampling in which a sampling interval width (K) is determined based on the needed sample size, and then every Kth element is selected from a sampling frame.

taxonomic analysis: Naturalistic data analysis technique in which the researcher organizes similar or related categories into larger categories and identifies differences between sets of subcategories and larger or overarching categories.

test-retest reliability: A reliability test in which the same test is given twice to the same subject under the same circumstances.

theoretical sensitivity: The researcher's sensitivity and ability to detect and give meaning to data.

theory: A set of interrelated constructs, definitions, and propositions that present a systematic view of phenomena by specifying relations among variables, with the purpose of explaining or predicting phenomena (see citation in Chapter 1).

trend study: Research examining a general population over time to see changes or trends that emerge as a consequence of time.

triangulation: The use of multiple strategies or methods as a means to strengthen the credibility of one's findings related to the phenomenon under study.

true experimental design: The classic two-group design in which subjects are randomly selected and randomly assigned (R) to either an experimental or control group condition. Before the experimental condition, all subjects are pretested or observed on a dependent measure (O). In the experimental group, the independent variable or experimental condition is imposed (X), and it is withheld in the control group. Subjects are then post-tested or observed on the dependent variable (O) after the experimental condition.

t-test: Parametric test to ascertain the extent to which any significant differences existing between two sample group means can be inferred to the population.

type I error: Rejecting the null hypothesis when it is true.

type II error: Failing to reject the null hypothesis when it is false.

univariate statistics: Descriptive data reduction approaches for one variable.

unobtrusive methodology (or nonreactive methodology):   Observation and examination of documents and objects that bear on the phenomenon of interest.

validity: The extent to which one's findings are accurate or reflect the underlying purpose of the study.

variable: A concept or construct to which a numeric value is assigned. By definition, a variable must have more than one value even if the investigator is interested in only one condition.

volunteer sampling: See convenience sampling.

Bibliography

Adams GR, Schvansveldt JD: *Understanding research methods,* ed 2, New York, 1991, Longman.

AEA Health Evaluation Newsletter, 1990, pp 1–2.

Agar MH: *The professional stranger: an informal introduction to ethnography,* New York, 1980, Academic Press.

Agar MH: Speaking of ethnography. In *Qualitative research methods,* vol 2, Newbury Park, Calif, 1986, Sage.

Agar MH: *Thinking of ethnography,* Newbury Park, Calif, 1986, Sage.

American Psychiatric Association: *Diagnostic and statistical manual (DSMIII-R),* ed 3, revised, Washington, DC, 1988, American Psychiatric Press.

Andrews FM, Klem L, Davidson TN, et al: A guide for selecting statistical techniques for analyzing social science data, ed 2, Ann Arbor, Mich, 1981, University of Michigan.

Argyris AC, Schon DA: Participatory action research and action science compared: a commentary, in Whyte WF, editor: *Participatory action research,* Newbury Park, Calif, 1991, Sage, pp 85–96.

Atchison DJ, Beard BJ, Lester LB: Occupational therapy personnel and AIDS: attitudes, knowledge and fears, *Am J Occupa Ther* 44:212–217, 1990.

Babbie E: *The practice of social research,* ed 6, Belmont, Calif, 1991, Wadsworth, p 118.

Babbie E: *The practice of social research,* Belmont, Calif, 1992, Wadsworth.

Bailey D: *Research for the health professional: a practical guide,* Philadelphia, 1991, FA Davis.

Bailey DM: Reasons for attrition from occupational therapy, *Am J Occupat Ther* 44:23–29, 1990.

Bellack AS, Hersen X, Michel X, et al: Single-case experimental designs. In *International handbook of behavior modification and therapy,* New York, 1982, Plenum Press, pp 167–203.

Bluebond-Langner M, Perkel D, Goertzel T, et al: Children's knowledge of cancer and its treatment: impact of an oncology camp experience, *J Pediatr* 116:207–213, 1990.

Brewer J, Hunter A: *Multimethod research: a synthesis of styles,* Newbury Park, Calif, 1989, Sage.

Brink PJ, Wood MJ: *Basic steps in planning nursing research: from question to proposal,* ed 3, Boston, 1988, Jones & Bartlett.

Burke JP, DePoy E: Emerging view of mastery, excellence and leadership in occupational therapy practice, *Am J Occupat Ther* 45:1027–1032, 1991.

Burns N, Grove S: *The practice of nursing research: conduct, critique and utilization,* Philadelphia, 1987, WB Saunders.

Butler S: Homeless women, unpublished dissertation, University of Washington, Seattle. 1991.

Campbell D, Fiske D: Convergent and discriminant validation by the multitrait-multimethod matrix, *Psychol Bull* 56:81–104, 1959.

Campbell DT, Stanley JC: *Experimental and quasi-experimental design,* Chicago, 1963, Rand McNally.

Case-Smith J: An efficacy study of occupational therapy with high-risk neonates, *Am J Occupat Ther* 42:499–506, 1988.

Cohen MR, Nagel E: *An introduction to logic and scientific method,* New York, 1934, Harcourt, Brace.

Cook TD, Campbell DT: *Quasi-experimentation: design and analysis issues for field settings,* Boston, 1979, Houghton Mifflin.

Cooper HM: *Integrating research: a guide for literature reviews,* ed 2, Newbury Park, Calif, 1987, Sage.

Corcoran MA, Gitlin LN: Environmental influences on behavior of the elderly with dementia: principles for intervention in the home, *Phys Occupat Ther Geriatr* 4:5–22, 1991.

Cox RC, West WL: *Fundamentals of research for health professionals,* ed 2, Rockville, Md, 1986, American Occupational Therapy Foundation.

Currier DP: *Elements of research in physical therapy,* Baltimore, 1990, Williams & Wilkins.

Currier DP: *Elements of research in physical therapy,* Baltimore, 1984, Williams & Wilkins.

Darroch V, Silvers FJ: *Interpretive human studies: an introduction to phenomenological research,* Washington, DC, 1982, University of American Press.

Davidoff GN, Keren O, Ring H, et al: Acute stroke patients: long-term effects of rehabilitation and maintenance of gains, *Arch Phys Med Rehabil* 72:869–873, 1991.

Denzin N: *Interpretive interactionism,* Newbury Park, Calif, 1988, Sage.

DePoy E: Mastery in clinical occupational therapy, *Am J Occupat Ther* 44:415–420, 1991.

DePoy E, Archer L: Quality of life in a nursing home, *Topics Geriatr Rehabil* 7:64–74, 1992.

DePoy E, Gallagher C, Calahoun L, et al: Altruistic activity v. self-focused activity: a pilot study, *Topics Geriatr Rehabil* 4:23–30, 1989.

DePoy E, Merrill SC: Value acquisition in an occupational therapy curriculum, *Occupat Ther J Res* 8:259–274, 1988.

DeVellis RF: *Scale development: theory and applications,* Newbury Park, Calif, 1991, Sage.

Duffy M: Methodological triangulation: a vehicle for merging quantitative and qualitative research methods, *Image J Nurs Scholarship* 19:130–133, 1987.

Douglas J, editor: *Understanding everyday life,* London, 1970, Routelage and Kegan Paul.

Estes J, Deyer C, Hansen R, et al: Influence of occupational therapy curricula on students' attitudes toward persons with disabilities, *Am J Occupat Ther* 45:156–165, 1991.

Evolving methodology in disability *Rehabil Briefs* XII:5, 1989.

Farber M: *The aims of phenomenology: the motives, methods, and impact of Husserl's thought,* New York, 1966, Harper & Row.

Fetterman DL: Ethnography, step-by-step. *Applied Social Research Methods Series,* vol 17, Newbury Park, Calif, 1989, Sage, pp 108–109.

Fetterman DL: A walk through the wilderness: learning to find your way. In Shaffir W, Stebbins R, editors: *Experiencing fieldwork: an inside view of qualitative research,* Newbury Park, Calif, 1991, Sage, pp 87–96.

Fielding N, Fielding J: *Linking data,* Beverly Hills, Calif, 1986, Sage.

Findley TM: II. The conceptual review of the literature or how to read more articles than you ever want to see in your entire LIFE, *Am J Phys Med Rehabil* 68:97–102, 1989.

Fortune AE, Hanks LL: Gender inequities in early social work careers, *Soc Work* 33:221–226, 1988.

Foshee V, Bauman KF: Parental and peer characteristics as modifiers of the bond-behavior relationship: an elaboration of control theory, *J Health Soc Behav* 33:66–76, 1992.

Frank G: Life history model of adaptation to disability: the case of a congenital amputee, *Soc Sci Med* 19:639–645, 1984.

French W, Bell, CH Jr: *Organization development, behavioral sciences interventions for organization improvement,* Englewood Cliffs, NJ, 1984, Prentice-Hall.

Geertz C: *The interpretation of cultures: selected essays,* New York, 1973, Basic Books.

Giannini EH: The N of 1 trials design in the rheumatic diseases, *Arthritis Care Res* 1:109–115, 1988.

Gilligan C: *In a difference voice: psychological theory and women's development,* Cambridge, Mass, 1982, Harvard University Press.

Gitlin LN, Burgh D, Durham D: Factors influencing therapist selection of adaptive technology for older adults in rehabilitation. In *Legacies and lifestyles for mature adults,* Philadelphia, 1992, Temple University.

Gitlin LN, Corcoran MA: Environmental modification to manage difficult behavior: a home intervention for caregivers; *Final report,* Rockville, Md, 1991, American Occupational Therapy Foundation.

Gitlin LN, Corcoran MA: Expanding caregiver ability to use environmental solutions for problems of bathing and incontinence in the elderly with dementia, *Technol Disabil* 2:12–21, 1993 .

Gitlin LN, Lawton MP, Landsberg LW, et al: In search of psychological benefits: exercise in healthy older adults, *J Aging Health* 4:174–192, 1992.

Gitlin LN, Levine R, Geiger C: Adaptive device use in the home by older adults with mixed disabilities, *Arch Phys Med Rehabil* 74:149–152, 1993.

Glaser B, Strauss A: *The discovery of grounded theory,* Chicago, 1967, Aldine.

Glaser B, Strauss A: *Grounded theory: strategies for qualitative research,* New York, 1967, Aldine.

Glaser BG: *Theoretical sensitivity: advanced in the methodology of grounded theory,* Mill Valley, Calif, 1978, Sociology Press.

Goodman C: Evaluation of a model self help telephone program: Impact of natural networks, *Soc Work* 35:556–562, 1990.

Grinnell RM: *Social work research and evaluation,* Itasca, Ill, 1988, Peacock.

Guba EC, editor: *The paradigm dialogue,* Newbury Park, Calif, 1990, Sage.

Guba EG: Criteria for assessing the trustworthiness of naturalistic inquiries, *Educ Commun Technol J* 29:75–91, 1981.

Gubrium J: Analyzing field reality. In *Qualitative research methods,* vol 8, Newbury Park, Calif, 1988, Sage.

Gubrium J: Recognizing and analyzing local cultures. In Shaffir W, Stebbins R, editors: *Experiencing fieldwork: an inside view of qualitative research,* Newbury Park, Calif, 1991, Sage, pp 131–142.

Hacker B: Single subject research strategies in occupational therapy, part 1 (research methods, case study), *Am J Occup Ther* 34:103–108, 1980.

Hacker B: Single subject research strategies in occupational therapy, part 2 (case study, research method), *Am J Occup Ther* 34:169–175, 1980.

Hasselkus B: Meaning in family caregiving: perspectives on caregiver/professional relationships, *Gerontologist* 28:686–690, 1988.

Hasselkus BR: The meaning of activity: day care for persons with Alzheimer disease, *Am J Occup Ther* 46:199–206, 1992.

Headland TN, Pike KL, Harris M: *Emics and etics: the insider/outsider debate,* Newbury Park, Calif, 1990, Sage.

Healthy people 2000, Washington, DC, 1990, US Government Printing Office.

Henkel RE: *Tests of significance,* Newbury Park, Calif, 1976, Sage.

Hirschi T: *Causes of delinquency,* Berkeley, Calif, 1969, University California Press.

Hodge R, Kress G: *Social semiotics,* Ithaca, NY, 1988, Cornell University Press.

Hoover KR: *The elements of social scientific thinking,* ed 2, New York, St Martin's Press.

Huck SW, Cormier WH, Bounds WG: *Reading statistics and research,* New York, 1974, Harper & Row.

Hume D: In Bigg LAS, editor: *A treatise on human nature,* New York, Clarendon Press.

Hurd P, Okolo EN, Hartzema AG, et al: Research design and data collection methods. In Okolo EN, editor: *Health research design and methodology,* Boca Raton, Fla, 1990, CRC Press, pp 97–146.

Husserl E [Boyce WR, translator]: *Ideas: general introduction to pure phenomenology,* New York, 1931, Macmillan.

Kachigan SK: *Statistical analysis: an interdisciplinary introduction to univariate and multivariate methods,* New York, 1986, Radius Press.

Kaplan S: Franco-American culture: knowledge by service providers, thesis, 1992, Smith School for Social Work, Mass.

Kerlinger FN: *Foundations of behavioral research,* New York, 1973, Holt, Rinehart & Winston.

Kerlinger FN: *Foundations of behavioral research,* ed 3, New York, 1986, Holt, Rinehart & Winston.

Knoke D, Bohrnstedt GW: *Basic social statistics,* Itasca, Ill, 1991, Peacock.

Kohlberg L: *The psychology of moral development,* New York, 1984, Harper & Row.

Krefting L: Reintegration into the community after head injury: the results of an ethnographic study, *Occup Ther J Res* 9:67–83, 1989.

Krueger R: *Focus groups: a practical guide for applied research,* Newbury Park, Calif, 1988, Sage.

Langness LL, Frank G: *Lives: an anthropological approach to biography,* Novato, Calif, 1991, Chandler & Sharp.

Levi-Strauss C [Jacobson C, Schoepf BG, translators]: *Structural anthropology,* New York, 1960, Basic Books.

Liebow E: *Tally's corner,* Boston, 1967, Little, Brown.

Lincoln YS, Guba EG: *Naturalistic inquiry,* Newbury Park, Calif, 1985, Sage.

Lincoln YS, Guba EG: But is it rigorous? Trustworthiness and authenticity in naturalistic evaluation. In Williams DD, editor: *Naturalistic evaluation,* San Francisco, 1986, Jossey-Bass.

Lofland J, Lofland L: *Analyzing social settings: a guide to qualitative observation and analysis,* ed 2, Belmont, Calif, 1984, Wadsworth.

Marshall C, Rossman GB: *Designing qualitative research,* Newbury Park, Calif, 1989, Sage.

Maruyama M: Endogenous research: the prison project. In Reason P, Rowan J, editors: *Human inquiry: a sourcebook for new paradigm research,* New York, 1981, John Wiley & Sons, pp 267–282.

Matza D: *Becoming deviant,* Englewood Cliffs, NJ, 1969, Prentice-Hall, p 25.

McCraken G: *The long interview,* Newbury Park, Calif, 1988, Sage.

Mechanic D: Medical sociology: some tensions among theory, method and substance, *J Health Soc Behav* 30:147–160, 1989.

Merton RK, Fiske M, Kendall P: *The focused interview: a manual of problems and procedures,* New York, 1956, Free Press.

Miles MB, Huberman AM: *Qualitative data analysis: a source book of new methods,* Newbury Park, Calif, 1984, Sage.

Miller D: *Handbook of research design and social measurement,* Newbury Park, Calif, 1991, Sage.

Mishel M, Murdaugh C: Family adjustment to heart transplantation: redesigning the dream, *Nurs Res* 36:332–338, 1987.

Mohr L: *Understanding significance testing,* Newbury Park, Calif, 1990, Sage.

Morgan DL: *Focus groups as qualitative research,* series no 16, Newbury Park, Calif, 1988, Sage.

Morse JM: *Qualitative nursing research,* Rockville, Md, 1989, Aspen.

Moustakas C: Heuristic research. In Reason P, Rowan J, editors: *Human inquiry: a sourcebook for new paradigm research,* New York, 1981, John Wiley & Sons, pp 207–218.

Nencel L, Pels P: *Constructing knowledge: authority and critique in social science,* Newbury Park, Calif, 1991, Sage.

Olson DH, Portner J, Lavee Y: FACES-III. In *Family Social Science,* Minneapolis, 1985, University of Minnesota.

Orcutt BA: *Science and inquiry in social work practice,* New York, 1990, Columbia University Press.

Osgood CE, Suci GJ, Tannenbaum PH: *The measurement of meaning,* Urbana, Ill, 1957, University of Illinois Press.

Ottenbacher K: *Evaluating clinical change: strategies for occupational and physical therapists,* Baltimore, Md, 1986, Williams & Wilkins.

Ottenbacker KJ, Bonder B: Scientific inquiry: design and analysis issues in occupational therapy. In American Occupational Therapy Foundation, Rockville, Md, 1986, p 196.

Oyster CK, Hanten WP, Llorens LA: *Introduction to research: a guide for the health science professional,* Philadelphia, 1987, JB Lippincott.

Palmore EB: Predictors of outcome in nursing homes, *J Appl Gerontol* 9:1172–1184, 1990.

Parse RR, Coyne AB, Smith MJ: Nursing research traditions: quantitative and qualitative approaches. In *Nursing research: qualitative methods,* Bowie, Md, 1985, Brady Communications, pp 1–8.

Patton M: *Qualitative evaluation and research methods,* ed 2, Newbury Park, Calif, 1990, Sage.

Payton OD: *Research, the validation of clinical practice,* Philadelphia, 1988, FA Davis.

Pearlin L: Structure and meaning in medical sociology, *J Health Soc Behav* 33:1–9, 1992.

Peirce CS: *Essays in the philosophy of science,* Indianapolis, Ind, 1957, Bobbs-Merrill.

Peller S: *Quantitative research in human biology and medicine,* Baltimore, 1967, Williams & Wilkins.

Perrow C: *Complex organizations: a critical essay,* ed 2, New York, 1979, Random House.

Piaget J [Warden M, translator]: *Judgment and reasoning in the child,* New York, 1926, Harcourt, Brace.

Pilcher D: *Data analysis for the helping professionals: a practical guide,* Newbury Park, Calif, 1990, Sage.

Polansky NA: *Social work research: methods for the helping professions,* Chicago, 1975, University of Chicago Press.

Posner J, Gorman KM, Gitlin LN, et al: Effects of exercise training in the elderly on the occurrence and time to onset of cardiovascular diagnoses, *J Am Geriatr Soc* 38:205–210, 1990.

Rabin AI, Zucker RA, Emmons RA, et al: *Studying persons and lives,* New York, 1990, New York Springer-Verlag, p IX.

Radloff LS: The CES-D scale: a self report depression scale for research in the general population, *Appl Psychol Measure* 1:385–401, 1977.

Randall JH, Buchler J: *Philosophy: an introduction,* New York, Barnes & Noble.

Reason P, Rowan J, editors: *Human inquiry: a sourcebook for new paradigm research,* New York, 1981, John Wiley & Sons.

Reid WJ, Smith AD: *Research in social work,* ed 2, New York, 1989, Columbia University Press.

Richardson L: *Writing strategies: reaching diverse audiences,* Newbury Park, Calif, 1990, Sage.

Roethlisberger FJ: *Management and morale,* Cambridge, Mass, 1941, Harvard University Press.

Roethlisberger FJ, Dickson WJ: *Management and the worker,* Cambridge, Mass, 1947, Harvard University Press.

Rossman GB, Wilson BL: Numbers and words: combining quantitative and qualitative methods in a single large-scale evaluation study, *Evaluation Rev* 9:627–643, 1985.

Rowles GD, Reinharz S: Qualitative gerontology: themes and challenges. In Reinharz S, Rowles GD, editors: *Qualitative gerontology,* New York, 1988, New York Springer-Verlag, pp 3–33.

Royeen C, editor: *Clinical research handbook,* Thorofare, NJ, 1989, Slack.

Rozier C, Gilkeson G, Hamilton BL: Job satisfaction of occupational therapy faculty, *Am J Occup Ther* 45:160–165, 1991.

Ruby J: Exposing yourself: reflexivity, anthropology and film, *Semiotica* 30:153–179, 1980.

Runcie JF: *Experiencing social research,* Homewood, Ill, 1976, Dorsey Press.

Russell H: *Introduction to mathematical philosophy,* New York, 1919, Macmillan.

Salomon G: Transcending the qualitative-quantitative debate: the analytic and systematic approaches to educational research, *Educ Research* 20:10–18, 1991.

Santiago JM, Bachrack LL, Berren MR, et al: Defining the homeless mentally ill: a methodological note, *Hosp Community Psychiatry* 39:1100–1102, 1989.

Savishinsky JS: *The ends of time: life and work in a nursing home,* New York, 1991, Bergen & Garvey.

Schon D: *The reflective practitioner,* New York, 1983, Basic Books.

Shaffir WB, Stebbins RA, editors: *Experiencing fieldwork: an inside view of qualitative research,* Newbury Park, Calif, 1991, Sage.

Smith AG, Louis KS: Multi-method policy research: issues and applications, *Am Behav Sci* 26:1–144, 1982.

Smith JK: Quantitative versus qualitative research: an attempt to clarify the issue, *Educ Researcher* 12:6–13, 1983.

Spanier GB: Measuring dyadic adjustment: new scales for assessing the quality of marriage and similar dyads, *J Marriage Fam* 38:15–28, 1976.

Sperber D: *On anthropological knowledge,* New York, 1987, Cambridge University Press.

Spielberger DC, Jacobs G, Russel S, et al: Assessment of anger: the state-trait anger scale, *Adv Personality* 2:159–187, 1983.

Spradley J: *Participant observation,* New York, 1980, Holt, Rinehart & Winston.

Spradley J, McCurdey D: *The cultural experience: ethnography in complex society,* Chicago, 1972, Science Research Associates.

Spradley JP: *The ethnographic interview,* New York, 1979, Holt, Rinehart & Winston.

Stearns CA: Physicians in restraints: HMO gatekeepers and their perceptions of demanding patients, *Qualitative Health Res* 1:326–348, 1991.

Steier F, editor: *Research and reflexivity,* Newbury, Calif, 1991, Sage.

Stein F: *Anatomy of clinical research: an introduction to scientific inquiry in medicine, rehabilitation and related health professions,* Thorofare, NJ, 1989, Slack.

Stewart DW: *Secondary research: information sources and methods,* Newbury Park, Calif, 1984, Sage.

Stiles WB, Snow JS: Counseling session impact as seen by novice counselors and their clients, *J Counseling Psychol* 31:3–12.

Stouffer SA: *Communism, conformity and civil liberties,* New York, 1955, Doubleday, pp 263–265.

Strauss A: *Qualitative analysis for social scientists,* New York, 1987, Cambridge University Press.

Subjects, respondents, informants or participants [editorial], *Qualitative Health Res* 1:403–406, 1991.

Strauss AI, Corbin J: *Basics of qualitative research: grounded theory procedures and techniques,* Newbury Park, Calif, 1990, Sage.

Suffet F, Lifshitz M: Women addicts and the threat of AIDS, *Qualitative Health Res* 1:51–79, 1991.

Tierney W: *Culture and ideology in higher education,* New York, 1991, Praeger.

Toseland RW, Rossiter CM, Smith GC: Comparative effectiveness of individual and group interventions to support family caregivers, *Soc Work* 35:209–217, 1990.

Tripodi T, Follin P, Meyer H: *The assessment of social research,* Trasca, Ill, 1983, Peacock.

Van Maanen J, editor: *Qualitative methodology,* Newbury Park, Calif, 1983, Sage.

Van Manen M: *Researching lived experience: human science for an action sensitive pedagogy,* Albany, NY, 1990, State University of New York Press.

Wax M: On misunderstanding verstehen: a reply to Abel, *Sociol Soc Res* 51:323–333, 1967.

Webb EJ, Campbell DT, Schwartz RD, et al: *Nonreactive measures in the social sciences,* 1981, Boston, Houghton Mifflin.

Wechsler D: *The measurement of adult intelligence,* ed 4, New York, 1958, Psychological Corp.

Wechsler D: *Wechsler adult intelligence scale–revised manual,* New York, 1981, Psychological Corp.

Whyte WF, editor: *Participatory action research,* Newbury Park, Calif, 1991, Sage.

Willoughby J, Keating N: Being in control: the process of caring for a relative with Alzheimer's disease, *Qualitative Health Res* 1:27–50, 1991.

Wilson HS, Hutchinson SA: Triangulation of qualitative methods: heideggerian hermeneutics and grounded theory, *Qualitative Health Res* 1:263–276, 1991.

Wilson J: *Thinking with concepts,* New York, 1966, Cambridge University Press.

Wuerst J, Stern PN: Empowerment in primary health care: the challenge for nurses, *Qualitative Health Res* 1:80–99, 1991.

Yerxa E: Seeing a relevant, ethical and realistic way to knowing for occupational therapy, *Am J Occup Ther* 45:199, 1991.

Yin RK: *Case study research,* Newbury Park, Calif, 1989, Sage.

Zarit SH: Do we need another stress and caregiver study? *Gerontologist* 29:47, 1989.

Zemke R: The continua of scientific research designs, *Am J Occup Ther* 43:551–553, 1989.

Zung WD: A rating instrument for anxiety disorders, *Psychosomatics* 12:371–379, 1971.

INDEX

INTRODUCTION TO RESEARCH
Multiple Strategies for Health and Human Services